Modelling Urban Development with Geographical Information Systems and Cellular Automata

Modelling Urban Development with Geographical Information Systems and Cellular Automata

Yan Liu

CRC Press
Taylor & Francis Group
Boca Raton London New York

CRC Press is an imprint of the
Taylor & Francis Group, an **informa** business

CRC Press
Taylor & Francis Group
6000 Broken Sound Parkway NW, Suite 300
Boca Raton, FL 33487-2742

First issued in paperback 2020

ISBN-13: 978-0-367-57743-8 (pbk)
ISBN-13: 978-1-4200-5989-2 (hbk)

This book contains information obtained from authentic and highly regarded sources. Reasonable efforts have been made to publish reliable data and information, but the author and publisher cannot assume responsibility for the validity of all materials or the consequences of their use. The authors and publishers have attempted to trace the copyright holders of all material reproduced in this publication and apologize to copyright holders if permission to publish in this form has not been obtained. If any copyright material has not been acknowledged please write and let us know so we may rectify in any future reprint.

Library of Congress Cataloging-in-Publication Data

Liu, Yan, 1965 Oct. 29-
   Modelling urban development with geographical information systems and cellular automata / Yan Liu.
      p. cm.
   Includes bibliographical references and index.
   ISBN-13: 978-1-4200-5989-2 (alk. paper)
   ISBN-10: 1-4200-5989-0 (alk. paper)
   1. City planning--Simulation methods. 2. Cities and towns--Simulation methods. 3. Cities and towns--Growth--Simulation methods. 4. Geographic information systems. I. Title.

HT166.L58 2009
307.1'21601'1--dc22   2008026910

Visit the Taylor & Francis Web site at
http://www.taylorandfrancis.com

and the CRC Press Web site at
http://www.crcpress.com

Contents

Preface...xi
The Author xiii

Chapter 1   Introduction to Urban Development Modelling 1

    1.1   Models and Modelling.. 2
        1.1.1   The Need for Models.. 2
        1.1.2   Characteristics of Models.. 3
        1.1.3   Types of Models 4
        1.1.4   Procedures of Model Building 6
        1.1.5   The Pitfalls 7
    1.2   Theoretical Approaches of Urban
        Development Modelling 7
        1.2.1   Urban Ecological Approach 9
        1.2.2   Social Physical Approach...................................... 10
        1.2.3   Neoclassical Approach 11
        1.2.4   Behavioural Approach.. 13
        1.2.5   Systems Approach 14
    1.3   Contemporary Practices of Urban
        Development Modelling 16
        1.3.1   Cities as Self-Organising Systems 16
        1.3.2   Fuzzy Set and Fuzzy Logic 19
        1.3.3   GIS and Urban Modelling 19
    1.4   Problems and Prospects... 20
        1.4.1   Theoretical Problems 20
        1.4.2   Technical Problems 22
        1.4.3   Future Prospects... 22
    1.5   Conclusion 23

Chapter 2   Cellular Automata and Its Application in Urban Modelling............. 25

    2.1   Cellular Automata Modelling... 25
        2.1.1   Cellular Automata Modelling: A Game 25
        2.1.2   A Simple Cellular Automata Model 27
                2.1.2.1   Five Basic Elements of
                        a Cellular Automaton................................ 28
                2.1.2.2   Mathematical Representation of
                        a Cellular Automaton................................ 29
        2.1.3   The Complex Features of Cellular Automata 29
    2.2   Cellular Automata in Urban Modelling............................ 30
        2.2.1   An Urban Cellular Automata 30

2.2.2   Advantages of Cellular Automata for Urban
Modelling 33
    2.2.2.1   Simplicity in Model Construction 34
    2.2.2.2   Modelling Spatial Dynamics to
Support "What If" Experiments................ 34
    2.2.2.3   A "Natural Affinity" with Raster GIS 35
2.2.3   Early Applications of Cellular Automata in
Urban Modelling 35
2.3   Contemporary Cellular Automata-Based Urban
Modelling Practices.. 38
2.3.1   Space Tessellation: From Regular to Irregular
Spatial Units 38
    2.3.1.1   Regular Cells of Small or
Large Resolution....................................... 38
    2.3.1.2   Using Irregular Spatial Units.................... 40
2.3.2   From Binary and Multiple to Continuous
Cell States.. 41
2.3.3   Neighbourhood Definitions..................................... 41
    2.3.3.1   "Action-at-a-Distance"
Neighbourhood 41
    2.3.3.2   Neighbourhood Size 42
    2.3.3.3   Neighbourhood Type 43
    2.3.3.4   Irregular Neighbourhood........................... 44
    2.3.3.5   Sensitivity Analysis 44
2.3.4   Variation in Transition Rules 45
    2.3.4.1   Constrained Cellular Automata 45
    2.3.4.2   The SLEUTH Model 46
    2.3.4.3   Fuzzy Constrained Cellular
Automata Models....................................... 47
    2.3.4.4   Transition Rules Derived from
Other Models 48
    2.3.4.5   Artificial Neural Network (ANN)-Based
Cellular Automata Models......................... 49
    2.3.4.6   Stochastic Cellular Automata Model.......... 50
2.3.5   Modelling Time... 51
2.4   Conclusion 51

Chapter 3   Developing a Fuzzy Constrained Cellular Automata
Model of Urban Development 53

3.1   Urban Development and Fuzzy Sets.................................... 53
3.1.1   Fuzzy Representation of Geographical
Boundaries.. 54
3.1.2   Fuzzy Set Theory 55
    3.1.2.1   Definition of Fuzzy Set............................. 55
    3.1.2.2   Membership Function 56
    3.1.2.3   Fuzzy Operation 58

3.1.3   Urban Development as a Fuzzy Process 59
    3.1.3.1   Defining Urban Areas 59
    3.1.3.2   Fuzzy Set Approach in Defining
             Urban Areas 60

3.2   Fuzzy Logic Control in Cellular Automata-Based
      Urban Modelling 62
      3.2.1   Linguistic Variables and Fuzzy Logic 63
          3.2.1.1   Linguistic Variables 63
          3.2.1.2   Basic Logic Terms and Reasoning 64
          3.2.1.3   Fuzzy Logic 66
      3.2.2   Fuzzy Logic Control 67
      3.2.3   Fuzzy Logic Control in Cellular Automata-Based
          Urban Modelling 69

3.3   Developing Fuzzy Constrained Cellular Automata for
      Urban Modelling 70
      3.3.1   The Temporal Process of Urban Development 70
      3.3.2   The Speed of Urban Development as
          a Fuzzy Set 73
      3.3.3   The Fuzzy Transition Rules and Inferencing 76
          3.3.3.1   Primary Transition Rules 76
          3.3.3.2   Rule Firing Threshold 77
          3.3.3.3   Secondary Transition Rules 79
          3.3.3.4   Rule Calibration 81
      3.3.4   The Defuzzification Process 82
3.4   Conclusion 83

Chapter 4   Sydney: Urban Development and Visualisation 85

4.1   Sydney's Urban Development and Planning 85
      4.1.1   Historical Threads of Development 88
      4.1.2   Urban Development and Planning 90
          4.1.2.1   County of Cumberland Planning
             Scheme (1948) 90
          4.1.2.2   Sydney Region Outline Plan (1968) 93
          4.1.2.3   Sydney into its Third Century (1988) 95
          4.1.2.4   Cities for the 21st Century (1995) 97
          4.1.2.5   City of Cities (2005) 97
      4.1.3   Issues Relating to Sydney's Urban Development 100
4.2   Data Collection and Processing 100
      4.2.1   Topographic Data 100
      4.2.2   Transportation Network 101
      4.2.3   Physical Urban Areas 102
      4.2.4   Land Excluded from Urban Development 102
      4.2.5   Urban Planning Schemes 103
4.3   Defining Sydney's Urban Areas with Fuzzy Set Theory 104
      4.3.1   Urban Area Criteria for Statistical Purposes 104

4.3.2   Defining a Fuzzy Boundary of Sydney's
Urban Areas 105

4.3.3   Visualising Sydney's Urban Development in
Space and Time 107

4.4   Conclusion 110

Chapter 5   Modelling the Urban Development of Sydney:
Model Specification, Calibration and Implementation 111

5.1   Model Specification 111

     5.1.1   Cell Size and State 111

     5.1.2   Neighbourhood Configuration 112

     5.1.3   Transition Rules 113

         5.1.3.1   Urban Natural Growth Controlled by
Primary Transition Rules 113

         5.1.3.2   Constrained Development by
Secondary Rules 114

         5.1.3.3   Flexibility in Rule Implementation 119

     5.1.4   The Temporal Dimension 120

5.2   Model Calibration 120

     5.2.1   Model Calibration Principles 120

     5.2.2   Simulation Accuracy Assessment 122

         5.2.2.1   The Error Matrix Approach 122

         5.2.2.2   A Modified Error Matrix Approach 124

         5.2.2.3   Kappa Coefficient Analysis 126

5.3   Model Implementation in GIS 128

     5.3.1   Cellular Automata Modelling and GIS 128

     5.3.2   The ArcGIS Approach 129

     5.3.3   Graphic User Interface Design 130

     5.3.4   Model Calibration 131

5.4   Conclusion 132

Chapter 6   Modelling the Urban Development of Sydney:
Results and Discussion 133

6.1   A Summary of Results from the Model 133

     6.1.1   The Simulation and Calibration Sequence
of the Model 133

     6.1.2   Overall Results under All Transition Rules 134

6.2   The Impact of Individual Factors on Sydney's Urban
Development 138

     6.2.1   Unconstrained Urban Growth 139

     6.2.2   Topographically Constrained Development 142

     6.2.3   Transportation-Supported Development 142

     6.2.4   Urban Planning Policies and Schemes 144

     6.2.5   Other Transition Rules 145

6.3 The Impact of Neighbourhood Scale on
the Model's Results.. 146
    6.3.1 Results from the Model under Different
Neighbourhood Scales... 146
    6.3.2 Simulation Accuracies of the Model over Time....... 149
    6.3.3 Neighbourhood Scale and Model Calibration.......... 151
6.4 Perspective Views on Sydney's Development to
the Year 2031 151
    6.4.1 Factors Affecting Sydney's Future Development..... 151
        6.4.1.1 Improvement in Transportation
Infrastructure... 152
        6.4.1.2 The Impact of the 2005 Metropolitan
Strategic Plan... 153
    6.4.2 Perspective Views of Urban Development under
Different Planning Control Factors......................... 153
6.5 Conclusion 157

Chapter 7 Future Research Directions... 159

7.1 Local and Global Transition Rules.. 160
7.2 Applications of Fuzzy Set and Fuzzy Logic 160
7.3 Urban Consolidation and Anti-urbanisation Processes......... 161
7.4 The Spatial Area Unit and Its Interaction with the
Neighbourhood Scale 162
7.5 Reapplicability of the Model 162

References 163
Index 177

Preface

Urban development and the migration of much of the population from rural to urban areas are significant global phenomena. Increasingly, more small isolated population centres are changing into large metropolitan cities at the expense of prime agricultural land and the destruction of natural landscape and public open space. This has attracted a lot of attention to the study of urban development under the theme of global environmental change. Various urban models have been built for this purpose. Amongst these, models based on the principles of cellular automata are developing most rapidly.

Urban development resembles the behaviour of a cellular automaton in many aspects. The space of an urban area can be regarded as a combination of a number of cells, each cell taking a finite set of possible states representing the extent of its urban development with the state of each cell evolving in discrete time steps according to some local transition rules.

In this book, a simulation model of urban development was developed based on the principles of the cellular automata. An innovative feature of the model is the incorporation of the fuzzy set and fuzzy logic approach. Instead of defining the state of cells as a binary mode of either non-urban or urban, urban development was regarded as a spatially and temporally continuous process. In this process, a cell might be in a non-urban (or rural) or a fully urban state, or it can also be in a state that is not rural/natural but yet not fully urbanised, that is, it is to some extent urbanised. Based on the fuzzy set theory, the extent to which a cell has undergone an urban development process can be represented by a fuzzy membership grade. Within this membership grade, a cell can be non-urban or fully urban with a membership grade of 0 or 1 respectively, or it can be at any stage of converting from non-urban to urban land use, in which case the membership grade is between 0 and 1 exclusively.

In addition to the use of the continuous cell states represented by the membership grade, fuzzy logic constrained rules were proposed to control the transition of cells from one state to another. With the fuzzy-logic-controller, the transition rules were defined not by deterministic numbers or explicit mathematical formulae but rather as "linguistic variables" representing a certain kind of precondition for a decision or a process. For instance, the development of an urban area is likely to be enhanced by *high* levels of accessibility, or constrained by an *unsuitable* topography. The application of the natural language statement is closer to the actual process of urban development whilst it also represents the uncertainty of various constraints on this development.

The fuzzy constrained cellular automata model of urban development was implemented in the ESRI's ArcGIS environment as an extension to its spatial analysis functions. It was applied to the metropolitan area of Sydney, Australia, to simulate the spatial and temporal processes of urban development from 1976 to 2031. A unique technique in the calibration of the model using the temporal data set of Sydney was presented. The model has not been developed simply as a predictive model. Rather, it functions as an analytical tool to evaluate the impacts of various factors—physical, socioeconomic, and institutional—on urban development. Through the implementation of various transition rules, the model generates different scenarios of urban development. Therefore, the model is useful for urban planners to answer "what if" questions.

There are seven chapters in this book. The first chapter provides a context of urban modelling and a theoretical as well as practical review of modelling techniques in urban development research. The second chapter introduces the cellular automata approach. Research on urban development based on the cellular automata approach is surveyed and the problems raised by using this approach are identified. Based on a thorough understanding of urban modelling and the applications of the cellular automata in this field, Chapter 3 develops a fuzzy constrained cellular automata model of urban development. These include the application of fuzzy set in defining urban states, and the identification of primary and secondary rules constrained by fuzzy logic in urban transitions.

Chapters 4 to 6 concern the application of the cellular automata model to simulate the process of urban development in metropolitan Sydney. In Chapter 4, the descriptions of Sydney in relation to urban development and planning are addressed, followed by the construction of a geographical database for the application of the cellular automata model developed in Chapter 3. The urban development of Sydney from the years 1976 to 2006 is visualised in a GIS. By using the Sydney database, the cellular automata model of urban development is tested and calibrated in Chapter 5. Through this testing and calibration, the model is used to understand Sydney's urban development in a cellular environment, and to evaluate the impact of various factors on Sydney's urban development. These factors include the physical constraint, transportation network, and urban planning in relation to various areas planned for urban development. By varying the size of the neighbourhood, the effects of different neighbourhood scales on the model's outcomes are analysed. In addition, options for the future urban development of Sydney under different planning control conditions in the next two and a half decades are predicted using the model.

Finally, conclusions are drawn in the last chapter on urban modelling using the cellular automata approach and the application of the model to simulate the actual process of urban development. With these concluding remarks, future research directions are mapped out, thus bringing the book to a closure.

The Author

Yan Liu is an assistant professor at the National Institute of Education, Nanyang Technological University, Singapore. She earned her B.Sc. in geography and her M.Sc. in economic geography from Hubei University and Central China Normal University, respectively, and her Ph.D. in geographical sciences from the University of Queensland, Australia. Prior to her current position she lectured at the Faculty of Engineering and Surveying of the University of Southern Queensland. Her research areas include urban modelling, geodemographics, and GIS in health and estate management as well as GIS in schools research.

1 Introduction to Urban Development Modelling

Urban development and the migration of population from rural to urban areas are significant global phenomena. Increasingly, more small, isolated population centres are changing into large metropolitan cities, with the conversion of natural land to urban use becoming quite obvious. According to the *2005 Revision of World Urbanisation Prospects* reported by the Department of Economic and Social Affairs' Population Division of the United Nations (United Nations 2006), in 1900, only 13% of the world's population lived in urban areas; this proportion increased to 29% by 1950, and it reached 49% in 2005. The latest U.N. population projection also indicates that the proportion of urban population will rise to 60% by 2030, which means that about 4.9 billion people out of a total world population of 8 billion are expected to be urban dwellers in 2030 (United Nations 2006).

The majority of urban growth will occur in the less developed countries. Although the patterns of urban growth in the developing countries are not of much difference to what happened in Europe and the United States a century ago, the absolute scale of this growth in terms of the number of cities undergoing rapid growth and the sheer number of people involved is much greater than ever before. According to the U.N. population projection, urban population growth in the less developed countries is projected to be 2.2% in average, annually, from 2005 to 2030, which is higher than the overall annual urban growth rate of 1.8% over the same period. As a consequence, the urban population in the less developed countries will increase from 2.3 billion in 2005 to 3.9 billion over the next 25 years (United Nations 2006).

In the more developed countries, the most rapid urban growth took place over a century ago, with the growth still continuing, although at a much slower rate on average than in previous decades. Much of the present population shift in the developed countries involves movement from the concentrated urban centres to vast, sprawling metropolitan regions or to small- and intermediate-sized cities, resulting in the physical expansion of the urban land and the conglomeration of multiple cities known as the megalopolis (Gottmann 1961).

Rapid urban development usually happens at the expense of prime agricultural land, with the destruction of natural landscape and public open space, which has an increasing impact on the global environmental change. As Vitousek (1994: 1861) states, "three of the well-documented global changes are increasing concentrations of carbon dioxide in the atmosphere, alterations in the biochemistry of the global nitrogen cycle, and an ongoing land-use/land cover change." The spatio temporal process of urban development and the social–environmental consequences of such development deserve serious study by urban geographers, planners, and policy makers because of the direct and profound impacts on human beings.

Cities are characterised by an immense complexity and internal heterogeneity (Bourne 1982). However, they also display a certain degree of internal organisation in terms of spatial patterns and temporal processes (Bourne 1982, 1971). Understanding the spatial patterns and temporal processes of urban development has been the subject of numerous historical studies involving the application of models.

This first chapter provides a brief discussion on models and model building in the context of urban development. It identifies and reviews major approaches conventionally applied in urban development modelling, and the strengths and weaknesses associated with each approach. Contemporary modelling practices under the self-organising paradigm are also reviewed, which are followed by discussions on the problems and prospects of urban development modelling.

1.1   MODELS AND MODELLING

In the Webster's Dictionary, a model is defined as "a description, a collection of statistical data, or an analogy used to help visualise often in a simplified way something that cannot be directly observed (as an atom)," or "a theoretical projection in detail of a possible system of human relationships" (Webster's 1964: 1451). In another dictionary, the *Collins English Dictionary,* a similar definition is given, which says that a model is "a simplified representation or description of a system or complex entity, especially one designed to facilitate calculations and predictions" (Makins 1995: 1003). These definitions show that a model in general is a simplified representation of reality. Modelling, therefore, is the process or behaviour of producing models. It also includes the act or art of those who produce models.

In geography, the terms *model* and *modelling* were given very broad interpretations in the 1960s. As employed in *Models in Geography* (Chorley and Haggett 1967), a model could be a theory, a law, a hypothesis, a structured idea, a role, a relation, an equation or a series of equations, a synthesis of data, a word, a map, a graph, or some type of computer or laboratory hardware arranged for experimental purposes. This broad definition of a model was narrowed by physical geographers to "any rule that generates output from inputs" (Haines-Young and Petch 1986: 145), or in other words, "any device or mechanism which generates a prediction" (Haines-Young and Petch 1986: 144). According to this definition, models are devices or mechanisms constructed on the basis of a theory that can generate new information to test the adequacy of the theory embedded in them. Haines-Young and Petch (1986) exclude models from theories, laws, or hypotheses, and they view modelling as "an activity that enables theories to be examined critically" (Haines-Young 1989: 22–23). However, in the area of human geography, the use of models was extended beyond the hypothetical–deductive view of science, which was "not necessarily positivist and functionalist simply because it is (often) a mathematical approach" (Wilson 1989: 64).

1.1.1   THE NEED FOR MODELS

The idea of using models in scientific research is by no means new. This idea comes from the way people react with the real world in which they live. Practically, all systems in the real world are exceedingly complex. Therefore, these systems are constantly explored by the use of simplified patterns of symbols, rules, and process (Apostel 1961; Meadows 1957). With the use of models, the complex systems of reality can be simplified so that they can be understood and managed.

The application of models in scientific research is important in many aspects. In one aspect, although it is accepted that models are different from theories, they play an important role in the development of theories. Models not only serve as a framework for theories to be expressed in a precise language, they also enable theories or hypotheses embedded in them to be examined. The relationship of modelling to the development of theory is extremely subtle and involves constant alternation between inductive and deductive reasoning. It is even argued that the extent of theoretical development in a field is partially equivalent to the extent to which it employs abstract models for analysis and prediction (Kilbridge, O'Block, and Teplitz 1970).

Models are also important in a practical context, especially when dealing with social systems that are often of concern to urban analysts and planners. Unlike laboratory scientists, urban analysts and planners can seldom manipulate the objects of their study to find the best arrangement or to discover natural properties or laws. The scales of cost and time are usually too large to allow for experimentation, and controlled experimentation with the social elements is rarely a possibility. In this case, through the construction of models, researchers can use them to represent the structure or function of the real system and to understand, explain, or predict the behaviour of the system. They can also use models to create an artificial environment for experimentation. For these reasons, models have been widely applied in urban planning, resources allocation, and prediction, as well as in assisting decision-making practices.

1.1.2 CHARACTERISTICS OF MODELS

As models are simplified structures of reality that present supposedly significant features or relationships in a generalised form, they do not include all the associated observations or measurements of the systems they model. Thus, the most fundamental feature of models is that their construction has involved a highly *selective* attitude to information. With this selective attitude, not only the noise but also the less important signals of the system have often been eliminated, enabling the fundamental, relevant, or interesting aspects of the real system to appear in some generalised form (Haggett and Chorley 1967). Therefore, models can be thought of as selective pictures of the real-world system, and "only by being unfaithful in some aspects can a model represent its original" (Black 1962: 220).

The selective feature of models also implies that models resemble the real-world system in some aspect; they are structured *approximations* of reality. A good model represents the real world in a simplified yet valid and adequate way. The model must be simple enough in order for one to easily understand and make decisions using it. It must be adequate to contain all the important elements of the real-world system, and it must be valid because all the elements modelled must be correctly interrelated according to their connections or structures.

Another feature of models is their *suggestive* nature, in that a successful model contains suggestions of its own extensions and generalisation (Hesse 1953).

For a predictive model, predictions about the real world can be made directly from the model. For a descriptive model, it can reveal much about the structure of the real-world system. Therefore, models provide concrete evidence of the way in which "everything affects everything else" (Lowry 1965: 159). It is because of this suggestive feature that models are widely applied in scientific research, especially in understanding social systems.

As models are generated from reality representing the fundamental features of that reality, they should be reapplicable to the real world (Chorley 1964). This implies that a model generated from one system should be applicable to other systems with similar features. In fact, many model builders have used the *reapplicability* of models to judge the value of the models in geographical research (Haggett and Chorley 1967).

1.1.3 TYPES OF MODELS

Models can be classified in various ways. In a general manner, they can be categorised into three types according to their degree of simplification and abstraction (Figure 1.1). The simplest level of abstraction occurs when reality is altered only in terms of scale. Scale models can be further differentiated as iconic models if they are miniature copies of reality, such as those employed by architects for the buildings they design, and as analogue models if the miniaturisation is accompanied by the transformation of certain properties such as the maps used by geographers.

At a greater level of abstraction are conceptual models in which the focus is upon relationships between different components of reality. The expression of this type of model can be in diagrammatic form or in verbal language. The von Thünen model of agricultural location is an example of this type (von Thünen 1826). Based on an econometric analysis of the estates in Mecklenburg, Germany, von Thünen modelled the land-use patterns resulting from the maximisation of rent at every site. His model was inherently descriptive (Henshall 1967). However, the framework of this model was used by later writers, notably Hoover (1936), Lösch (1943), and Dunn (1954), as a basis for normative models (Henshall 1967).

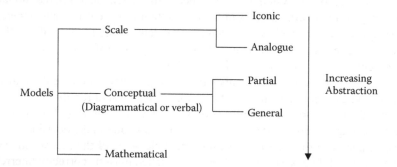

FIGURE 1.1 Types of models. (Adapted from Thomas, R. W. and Huggett, R. J., *Modelling in geography: a mathematical approach*, Barnes & Noble Books, New Jersey, 1980. With permission.)

Models at the highest level of abstraction are mathematical models; these are the most widely used and are often the major concern in scientific research. However, models in this category can be further divided into different subclasses. Table 1.1 summarises some of the classifications used frequently in urban modelling (Robinson 1998; Batty 1976; Kilbridge, O'Block, and Teplitz 1970; Haggett and Chorley 1967).

The cellular automata model of urban development that is central to this book is a simulation model based on the principles of nonlinear systems and the Chaos Theory. The model simulates the processes of urban development in space and time, which can be used to test various "what if" scenarios on urban development. It fits in the category of a theory-based model and is a general model in regard to the substantive issues being modelled. It is both descriptive and normative in nature and is dynamic over time. Through the process of model simulation, the outcomes of the model are generated in stages.

TABLE 1.1
Different Classifications of Models

Classification criteria	Type of models	Explanation of classification criteria
Theoretical base of models	Theory-based models	Models are derived directly from a theory as a symbolic statement of the theory.
	Theory-laden models	The real world phenomena are abstracted to symbolic forms and are related structurally in a model, thus to create new theory.
Substantive issues being modelled	Partial models	Models deal with only a part of the system being modelled or a subsystem of the reality.
	General models	Models attempting to deal with two or more subsystems of the reality being modelled.
Descriptive or normative features of models	Descriptive models	Descriptive models deal with some stylistic description of reality.
	Normative models	Normative models deal with what might be expected to occur under stated conditions.
The way models deal with time	Static models	Models concentrating on the equilibrium structural features.
	Dynamic models	Models concentrating on processes and functions through time.
The predictive nature of models	Deterministic models	Models are based on the notion of exact prediction, which is produced by natural and physical laws.
	Stochastic models	These are also called probabilistic models, which involve the use of probabilities, and they produce a range of possible outcomes rather than a single prediction.
The solution procedure of models	Analytic models	Analytic solution procedures are direct and do not involve any form of iteration.
	Simulation models	Solutions in these models are gradually reached in stages.

Except for the foregoing classifications, mathematical models can also be classified according to their objectives, the techniques in use, or the theories or hypotheses underpinning them. A review of conventional models of urban development based on their underlying theoretical approaches is presented in Section 1.2.

1.1.4 PROCEDURES OF MODEL BUILDING

Although models vary significantly from one type to another, they share common procedures in the process of building them. Figure 1.2 shows the various stages of a modelling process that has been well accepted by many model builders (Caldwell and Ram 1999).

According to the flow chart, the first stage of model construction is to be clear about the objectives and become familiar with the problem that needs to be solved in the real world. Then, based on the understanding of the real-world system and the objectives, the model builder can begin setting up a model for the second stage. This includes selecting an appropriate theoretical base and designing a logical framework to encompass the objectives. In the absence of a well-established theory, the model builder needs to make assumptions or hypotheses that serve as the model's theoretical foundation. Decisions concerning the selective features to be modelled or neglected also need to be made at this stage.

The third stage is to formulate the model in a mathematical language. This is the critical stage and often the most difficult. At this stage, information about the real-world situation needs to be abstracted and translated into equations or other mathematical statements. With this formulation, the fourth stage is to implement the model in a computer programme and to fit or calibrate the model vigorously. This task involves two types of transformations: (a) to give precise empirical definition to the variables used in the model, and (b) to provide numerical values for the model's parameters. For some kinds of models, these variables or parameters can be verbal

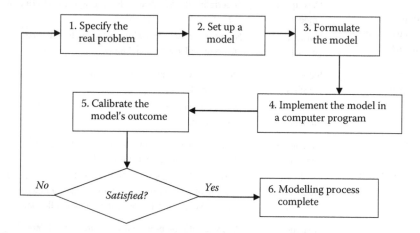

FIGURE 1.2 A flow chart showing the various stages of a modelling process. (Adapted from Caldwell, J. and Ram, Y. M., *Mathematical modelling: concepts and case studies,* 3rd ed., Kluwer Academic, Dordrecht, Netherlands, 1999. With permission.)

or fuzzy values. Once the variables and parameters are provided, the model can be coded in a computer programme and run.

At this stage, although the model has been established and outcomes can be achieved by running the model, a question remains about the final product: does it really work? Or, how good are the model's outcomes compared to reality? So, the task in the fifth stage is to test the model's outcomes. This includes testing the validity and sensitivity of the model to see if it behaves reasonably when changes are made to some of the conditions, and evaluating whether or to what extent the model resembles the real-world situation. If one feels that the model can be improved in any aspect, then one needs to go back to stage one and complete the modelling process again. In this respect, modelling can be regarded as an iterative process where one starts from a crude model and gradually refines it until it is good enough to solve the original problem. In addition, as the real world is constantly changing, models that once were acceptable may no longer be adequate or valid. Therefore, continual validation of the models is required (Dickey and Watts 1978).

1.1.5 THE PITFALLS

Although there is little doubting the value of using models, especially when dealing with complex systems, the characteristics of the models discussed in Section 1.1.2 imply that there are many dangers to which model builders may fall prey (Haggett and Chorley 1967: 26). As models are simplified abstractions of reality, they leave some parts of reality behind. This is both a strength and a weakness. The strength lies in the clarity of the essentials and the manipulative nature of symbols; the weakness lies in a certain necessary degree of invalidity. Also, simplification may lead to the risk of "throwing the baby out with the bath water" (Haggett and Chorley 1967: 26). The suggestive nature of models might lead to the improper use of prediction, structuring features to spurious correlations, and the approximation to unreality. Only when the limitations of the models are carefully borne in mind and the dangers of using models are remembered can one use models as an effective tool for solving scientific problems.

1.2 THEORETICAL APPROACHES OF URBAN DEVELOPMENT MODELLING

The use of models in urban research can be dated back to von Thünen's classical model of agricultural location (von Thünen 1826). In his famous book, *Der Isolierte Staat*, published in 1826, von Thünen considered the relationship of three factors: the distance of the farmers from the market, the prices received by the farmers for their goods, and the land rent. Based on an econometric analysis of the estates in Mecklenburg in north Germany, where von Thünen farmed for 40 years, from 1810 until his death in 1850, he hypothesized that the intensity of land use was inversely proportional to the transportation cost or distance from the market. In an "isolated state" with only one central city as the sole market and a uniform plain surrounding the city, this generates a concentric land-use pattern with the least intensive land use located the farthest away from the city centre (Henshall 1967).

Many models of urban development are related to von Thünen's model. For instance, Weber's (1909) *Industrial Location* could be regarded as the first model of urban growth. Later on, Christaller (1933) developed his *Central Place Theory* to model urban growth patterns in a regional context. With regard to the internal urban structure, Burgess' (1925) *Concentric Zone Model,* Hoyt's (1939) *Sector Model,* and Harris and Ullmans' (1945) *Multiple Nuclei Model* are three classical models of urban growth and urban land-use patterns (Hägerstrand 1967, 1965). These models were based on the understanding of urban development from the central business district (CBD) outwards. However, they are static, with little or no regard for the dynamic nature of urban development. Moreover, the models were based on assumptions that were far from practical; they were by no means operational.

The most widespread use of models in urban geography was during the period of the *quantitative revolution* in geography, which began in the late 1950s and continued till the late 1960s (Batty 1981). This development came almost exclusively from North America as a result of both the practical need and technical support of computer applications. In practice, the increasing car ownership during the 1940s and early 1950s led to the growing realisation that cities with their traditional physical forms could simply not cope with the "new mobility" (Batty 1976: 6). This led to the formation of transportation models in the late 1950s. The development of digital computing provided a means of working with complex mathematical models. During this period, a diverse array of styles, techniques, and applications of urban models were developed. The subjects of these models included land use, transportation, population, and urban economic activities, and were built and implemented using broad styles or techniques of linear analysis, mathematical programming, and simulation (Kilbridge, O'Block, and Teplitz 1970). These models were regarded by planners as providing artificial laboratories for experiments with urban structures (Dyckman 1963).

However, as the emphasis of these models was on the modelling techniques rather than their theoretical representations, they were seriously criticised by many researchers, thus resulting in an obvious shift of interests in the late 1970s from using mathematical models to qualitative analyses in urban research. This shift was maintained till the late 1980s when study on complex and open systems provided alternative ways to understand cities as evolutionary and complex systems (Allen 1997). The development of the geographical information system (GIS) and the integration of a GIS with urban modelling have also facilitated urban modelling with rich data sources and new techniques. These new developments have pushed the efforts of urban development modelling into a new era.

The rest of this section reviews various urban modelling approaches and practices in the literature. However, it is not the intention of the author to cover the vast volume of work undertaken in urban modelling as such an effort is beyond the scope of this book. Instead, the theoretical approaches underpinning these modelling efforts are summarised and an outline of the key themes of urban modelling is produced. It should be noted that although there are obvious differences between the various approaches discussed in the following text, they share some similarities. For example, all approaches sought to examine the patterns and processes of urban development, and they all regarded urban development as the outcome of the combination of human choices and actions and the wider social processes that placed constraints upon human actions (Hall 1998). It is the relative importance of the choices and constraints that these approaches emphasise, and the ways in which each is believed to operate, that distinguish the following approaches.

1.2.1 URBAN ECOLOGICAL APPROACH

This approach is based on the belief that human behaviour is determined by ecological principles, such as competition, selection, succession, and dominance. As in plant ecology, the most powerful human group would obtain the most advantageous position in a given urban environment, for example, the best residential location. This approach can be traced back to the work of the Chicago School of Human Ecology in the 1920s, and the most notable models of this approach were Burgess's (1925) concentric zone model, Hoyt's (1939) sector model, and Harris and Ullman's (1945) multiple nuclei model.

Burgess's model of urban growth was based on the notion that various elements of a heterogeneous and economically complex urban society actively compete for favourable locations within the city. The competition at the urban centre means a successive outward expansion of urban land, forming a series of concentric zones that encircle the centre (Carter 1995).

Although Burgess's model described an ideal pattern of urban growth, it did not take into account various urban environmental factors such as topography or transportation networks that can cause disturbances in the ideal pattern. This model was an oversimplification of reality and encouraged the postulation of other models of urban structure and growth.

Based on the studies of changes of residential patterns of 142 cities for the years 1910, 1915, and 1936, Hoyt (1939) advanced a sector model in which he identified that homogeneous areas of residence tended to grow outward from the centre toward the periphery in wedge-shaped sectors. In his sector model, in addition to the obvious emphasis on transportation routes where urban growth was often focused on, Hoyt also considered the effects of topographic variations and the adjacent and nearby land use on urban development. Although limited by the lack of theoretical exploration, Hoyt suggested that high-class residential areas might be expected to expand along established lines of travel in the direction of an existing nucleus of buildings. High-rent areas tended to spread along higher ground or along waterfronts (if those areas were not already occupied by manufacturing industries) and were also likely to grow toward the homes of the community leaders. The consideration of the various factors in the physical patterning of urban land use suggested a problem for the early ecologists' hope for developing a general model of urban land use. Such a general model was proven to be more difficult to achieve by Harris and Ullman (1945).

According to Harris and Ullman (1945), the patterns of urban growth and change still followed the general ecological principles identified by Burgess. For example, some activities always tend to be located in the vicinity of each other, and others repel each other, whereas some cannot afford the high rents demanded for the best sites. However, this growth was not centred around one single central business district but on certain growing points or "nuclei." These multiple nuclei attracted and repelled newcomers in the broadened urban areas. Each area was formed in consideration of multiple adjoining areas. As the city grew and changed, some new districts became more attractive than others. For example, heavy industry was likely to emerge in two or more outlying areas as the inner city became increasingly congested, new business districts were generated to serve populations some distance away from existing centres, and existing towns on the periphery were incorporated into the sprawling ecology of the growing metropolitan centres. The multiple nuclei model was "conceived as a further move away from the massive generalisation and toward reality" (Carter 1995: 132). However, it did not suggest any uniform patterning of land use among cities. In this respect, it would be appropriate to regard Harris and Ullman's model as a guide to thinking about the structure of cities rather than as a rigid generalisation about urban form.

Although classical models of urban ecology presented the general rules of urban growth and structure, these models were overly simplistic (Flanagan 1990). Thus, urban ecologists tried to refine the methodology to approach the complexity of the urban environment. With this concern in mind, Shevky and Williams (1949) developed a technique called "social area analysis," which was further formalised by Shevky and Bell (1955). In this technique, multivariate statistical analysis was used to identify three fundamental features to serve in various areas of the city, separate from one another. These features were economic status, family status, and ethnic classification. Through the analysis of these features, the urban area was structured as distinguishable social areas. This technique gained wide appeal among researchers in the 1960s. With the assistance of computer technology and the periodically available data from each census, factor analysis was further developed as a statistical tool. However, in the flood of the factorial ecology of different towns and cities, the theoretical content of this approach became "steadily diluted" (Bassett and Short 1989: 178). Whereas the overly simplistic models of classical urban ecology failed to describe much regarding existing patterns, the new method of factor analysis was seen as describing much but explaining little. Thus, after one decade of overindulgence, interest in this approach began to fade by the early 1970s.

1.2.2 SOCIAL PHYSICAL APPROACH

The social physical approach was based on the concept of human interaction in space. This approach was first developed as a direct analogy to physics. That is, it uses Newton's *Law of Gravitation* as an analogue for social interaction between places. It proposed that the movement of human activities such as changes in residence and employment between places were directly proportional to the mass of the activity at the origin and destination, and inversely proportional to the cost (in terms of distance or time) separating them. The model developed from this analogy was referred to as the gravity model, which was widely applied in studies of migrations, settlement network, and the intraurban structure in the 1960s. A lot of urban models under the paradigm of this social physical approach were developed by planners in connection with particular studies of individual cities or metropolitan areas, which often had direct operational capabilities.

In a typical gravity model, factors such as basic employment, economic structure, and population were usually distributed using particular allocation functions.

These functions were formulated based on spatial accessibility analysis. The gravity models had an obvious aptitude for prediction. A well-known and well-documented model using this approach was the one developed by the Chicago Area Transportation Study (CATS) for forecasting land uses in Chicago City up to the 1980s (Hambury and Sharkey 1961).

Following the extensive applications of the gravity models in urban spatial interaction studies, Wilson (1970) developed the social physical approach by introducing the second law of thermodynamics—the maximum entropy law—into this approach. Based on the principles of the maximum entropy law, Wilson formulated his entropy-maximising spatial interaction model. In this model, the movements of people and goods in cities were treated in the manner that particles in gases were treated in statistical mechanics using grand canonical ensembles and distinguishing them by origin and destination as "types" and by origin–destination pairs as "states" (Wilson 1984: 205). This is a macro-scale or aggregative approach, the success of which has been the ease of using it through aggregating the neoclassical models of consumers and producers (Robinson 1998).

Although the social physical approach was applied widely in urban planning models, the limitations of this approach are very clear. The fundamental limitation is that it fails to make an adequate representation of the behavioural process that leads to individuals selecting a particular journey to work. Models developed under this approach were aggregates; they stressed group behaviour rather than individual behaviour, and they lacked the ability to deal with important but non-quantitative factors. In addition, as the models were based on analogue assumptions to theories in physics, the theoretical base of this approach was very weak. As Batty (1982) argued, modellers were too concerned with technical issues rather than their theoretical foundations and policy implications.

1.2.3 NEOCLASSICAL APPROACH

The economic equilibrium or neoclassical approach was rooted in the tradition of economic theories. The von Thünen model of agricultural location can be regarded as the earliest model developed under this approach. Early works on this approach also included those of Weber (1909), Lösch (1943), and Isard (1956).

The neoclassical approach was built on the belief that the process of urban development is essentially an economic phenomenon, being driven by market mechanisms and the natural forces of competition among economic activities and social groups in an urban area. According to the economic theory of equilibrium, the allocation of urban land to various users in both quantitative and locational aspects is controlled by supply-and-demand relationships obeying the general rule of least costs and maximum benefits, or the utility maximisation rule in an equilibrium system. Under severe limiting assumptions, a typical model of urban economics shows urban structure as the reflection of spatial patterns of transport costs and urban land rent. The assumptions might be a concentric, homogeneous city with one single centre; the concentration of production of a composite consumption good; housing demand relating only to plot size, location, and externalities; and the ignorance of public sector policies. Examples of these models include Wingo's (1961) model of residential land development, and Alonso's (1964) and Lowry's (1964) models of urban structure.

Wingo (1961) was the first researcher who developed a concept of transportation demand considering the spatial relationship between home and work. With the journey to work viewed as the technological link between the labour force and the production process, he defined the demand for movement as the total employment of an urban area multiplied by the frequency of work. Drawing on a concept of accessibility, Wingo used a unit of measurement that was calculated as the cost of transportation based on the time spent in movement between points, or the out-of-pocket costs for these movements expressed in money equivalent to distance and number of trips. By substituting transportation costs for location costs, Wingo achieved a locational equilibrium for the distribution of households of particular rent-pay abilities to sites of a particular structure of rents. With the maximisation rule of utility, Wingo's concept generated a rent and residential density gradient that declines from the centre to the periphery.

Similar to Wingo's concept, Alonso's (1964) model also emphasised the substitution of transport inputs and lot size. He assumed that, in the land development process, individual decision-makers, be they firms or households, aimed to minimise rent and transportation costs and maximise the area occupied. Alonso defined a bid-rent curve in his model as a set of combinations of rent and transport inputs to represent an equal satisfaction level for an individual. If the rent curves for lower income groups were steeper for any pair of households with identical tastes, the poor would live at high densities near the city centre, whereas the rich would live at lower densities near the periphery. Both Wingo and Alonso confirmed Clark's (1951) original speculation on the distribution of population density gradients within a city and suggested that the negative exponential equation of population density distribution was a general equation that could be derived as a logical extension of the theory of urban land market.

Compared with the above two models, Lowry's (1964) model was probably the most widely applied under the economic equilibrium approach of the 1960s. His model was based on two underlying assumptions: one was that residential densities within a city fell away in a logical manner around the centres of employment, and the other was that the location and employment levels of the service sector were strongly influenced by accessibility to local customers whose effect was progressively reduced the farther away they lived. Under these assumptions, Lowry divided urban employment into basic and non-basic parts in an urban economic base. With the allocation of basic employment to various predetermined locations, the residential population that was likely to be associated with basic employment was calculated and allocated to the area. Service employment was then allocated in relation to the distribution of the residential population. To maintain a "rational" equilibrium of employment, and residential and service populations, the model was run repeatedly if non-basic employment happened in the area (Johnson 1972: 191). The model was comparatively simple for practical application. Therefore, this approach was widely accepted and applied in planning practices in the 1960s.

Although the standard neoclassical models were criticised both for their static equilibrium form and their simplified assumptions (Angel and Hyman 1972; Wheeler

1970; Boyce 1965), much of the work carried out in the 1970s was concerned with relaxing the various assumptions of this approach. Thus, models in this period incorporated multiple urban centres, different transport modes, externalities such as pollution, and public goods. Residential location models also incorporated income variations, differences in household preferences, variations in environmental quality, and racial discrimination in housing markets. However, because of its ignorance of the impact of human behaviour on urban growth and patterning process, this approach received fundamental criticism from behavioural approach theorists and the humanists.

1.2.4 BEHAVIOURAL APPROACH

Strictly speaking, this approach was referred to as the cognitive behavioural approach. It was developed from the criticism of the oversimplified concepts of human behaviour implicit in the urban ecological and neoclassical approaches. Unlike those positivist approaches that explored only a narrow aspect of human behaviour on utility maximisation, the behavioural approach sought to focus attention on the motivations behind individual behaviour, the way in which individuals search and learn about their urban environment, and their decision-making processes. This approach often involved explicit rejection of the assumption of the rational economic person and the simple utility-maximising framework (Johnston and Wrigley 1981). The urban development model developed at the University of North Carolina led by Stuart Chapin is a good example of this approach (Chapin and Weiss 1962a).

With the objective of approaching the dynamics of urban growth, the North Carolina Group developed a common framework with four elements for their studies. These four elements were the value system, behaviour patterns, urban development, and the control process. The central concern of this framework was the behaviour patterns that were the representations of human actions. Urban development was viewed as an end result of human actions, and the value system of urban society as the primary source of the impulse for actions. The objective of this framework was to seek explanations of urban development in terms of human behaviour, with the behaviour patterns being a function of people's values. The fourth element of this framework, the control process, concerns how influence could alter or affect behaviour patterns and thereby modify urban development toward certain predetermined goals. This element is often referred to as urban development strategies and plans. Under this framework, urban development was first viewed as the consequence of certain strategic decisions that structure the pattern of growth and development, and then as the consequence of the myriad of household, business and government decisions that followed from the first key decisions. Therefore, the focus of this framework was on decisions that were critical in the behavioural sequences of human actions. The identification, measurement, and interpretation of the value systems that formulate human behaviour were essential for the full analysis of urban development.

The application of this framework has several advantages, such as the emphasis on the decision-making process of urban development and the introduction of a control process. However, like other models developed under the behavioural approach, the model's focus was on simulating the individual decision-making process, in this case the distribution of households to available land. Although group decisions were regarded as key decisions in the conceptual framework, they were assumed to be known (Chapin and Weiss 1962b).

Because of its overemphasis on individual behaviour rather than group behaviour, and other weaknesses such as the overly simplistic view of the relationship between cognition and behaviour and the lack of general applicability, the behavioural approach was attacked from a variety of directions in the late 1970s. This provoked a rethinking by those who wanted to defend the basic concepts of the behavioural approach. Attempts were made both in strengthening the initial crude behavioural assumptions in fields such as innovation diffusion and residential mobility, and in building links between models of individual behaviour and wider societal constraints (Bassett and Short 1989).

1.2.5 SYSTEMS APPROACH

The systems approach was first used in urban modelling in the 1960s. It was based on the notions of the General Systems Theory. According to von Bertalanffy (1968), everything exists in a sort of system in which it becomes an element. All elements of the system are linked and interrelated and are also linked to the system's environment. For instance, an urban system consists of a set of elements or subsystems, such as population, land, employment, services and transport, to mention a few. All elements within the system are interacting with each other through social, economic, and spatial mechanisms while they are also interacting with elements in the environment. The significance of any one element does not depend on itself but on its relationships with others. It is the links between the different elements of the system that determine its evolution and so permit the process of change in the system. Thus, the focus of the systems approach is not on any single element but the connections and processes that link all the elements (Chisholm 1967). This suggests the application of systems analysis in dealing with the system.

The implementation of systems analysis involves two key steps: the first is the definition of a particular system as the object of study, and the second the way of describing the structure and behaviour of the system. In regard to the definition of systems, Chorley and Kennedy (1971) identified four different types of systems: the morphological system, the cascading system, the process–response system, and the control system (Figure 1.3). A morphological system represents the static relationships as links between elements, whereas links in a cascading system pass energy from one element to another. The process–response system combines the first two types of systems, but the focus of studies on this system is on the process rather than the form, with the emphasis on causal relationships. This system has two subtypes of simple action and feedback systems. The fourth type of system, that is, the control system, represents a special case of the process–response system, in which one or more elements act as valves to regulate the system's operation and therefore may be used to control it.

In order to illustrate the structure and behaviour of systems, a diverse range of mathematical methods has been employed. This includes factor analysis, principal component analysis, multicriteria analysis, linear and nonlinear programming, as

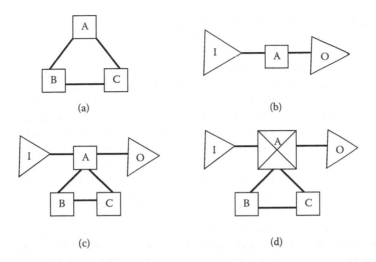

FIGURE 1.3 Portrayals of different systems: (a) morphological, (b) cascading, (c) process–response, and (d) control. A, B, and C are three systems; I = input to a system; O = output from a system. (From Robinson, G. M., *Methods and techniques in human geography*, John Wiley, New York, 1998. With permission.)

well as dynamic system simulation. Among these methods, the technique of Systems Dynamics developed by Forrester (1971, 1969, 1961) for simulating the urban development processes played an important role and hence is worth noting.

The Systems Dynamics technique was first developed for simulating industrial process in firms, but this technique has wide applicability (Batty 1976). Derived from the idea of control engineering, a system developed using the System Dynamics approach is closed with a feedback loop structure. The system is bounded to include those interesting components necessary to generate the modes of behaviour of interest. It is closed so that outside occurrences can be viewed "as random happenings that impinge on the system and do not themselves give the system its intrinsic growth and stability characteristics" (Forrester 1969: 12). The feedback loops are the fundamental building blocks of the system. The dynamic behaviour of the system is generated within feedback loops that are conceived in terms of levels of stocks (level variables). The level variables are progressively altered over time by the rate of change (rate variables). The rate variables are affected by various positive and negative feedback loops within the system. Through the use of a special computer language called DYNAMO designed by workers at the Massachusetts Institute of Technology (MIT), System Dynamics has been used to simulate the dynamic behaviour of urban systems and provide predictions on urban development under different conditions. This technique was regarded as providing a laboratory for strategic and tactical research (Forrester 1969). However, the lack of theoretical support and the difficulty in testing the relationships being modelled to any significant extent have caused criticism of this approach, and the non-spatial characteristics of this approach has also limited its application in modelling urban development in space and time.

The systems approach presented researchers with a way of constructing models beyond the simple cause–effect or stimulus–response relationships. Therefore, these models are widely accepted especially by planners in modelling the behaviour of urban systems and predicting future urban developments. However, the weakness of this approach is also obvious. For instance, in order to make models operational, some models were focused on small parts of the system they model rather than the system as a whole; others used simplified mathematical formula to represent the complex relationships or links between elements in the system. In dealing with the temporal change of systems, the systems approach freezes changes in systems rather than modelling the change itself. Therefore, the temporal change of a system is dealt with from an essentially static perspective (Robinson 1998). The limitations of the systems approach based on the General Systems Theory imply that the complexity of reality needs to be understood and modelled in new ways.

In the past three decades, studies of nonlinear process and open systems have led to the emergence of new understandings of complex systems and their evolution. Based on these understandings, cities are looked at as complex and open systems that have the capability of self-organisation. The concepts of self-organisation, chaos, and complexity theories have led to fruitful studies on urban development, which will be discussed in the following section.

1.3 CONTEMPORARY PRACTICES OF URBAN DEVELOPMENT MODELLING

Starting from the 1980s, the advances in nonlinear systems, fractals, and the Chaos Theory have led to new ways of looking at cities and their development, which led to significant progress in urban modelling both theoretically and practically. Theoretically, studies on nonlinear systems have led to the understanding of urban development as an irregular process in the manner of bifurcation and chaos (Allen 1997; Batty and Longley 1994; Wilson 1981a,b). Practically, the emergence of new digital data sources and GIS techniques have provided urban analysts with not only rich data sources but also new platforms and techniques for data management, analysis, and visualisation (Sui 1998; Nyerges 1995). In addition, the application of the fuzzy set theory and fuzzy logic in urban modelling has provided ways of soft computing that are closer to the real process of the progress of systems (Wu 1998b, 1996; Openshaw and Openshaw 1997).

1.3.1 CITIES AS SELF-ORGANISING SYSTEMS

Although the systems analysis based on the General Systems Theory provided an approach to look at cities in complex ways, it is essentially static and is limited in the ability to deal with dynamic growth phenomena. Following the General Systems Theory, studies on systems went beyond to focus on several special types of systems, including catastrophic and chaotic systems. In a catastrophic system, critical values of system parameters exist at which some unusual behaviour of the system can occur. Both the catastrophe and bifurcation theories are concerned with these critical values. Studies in the Chaos Theory reveal that even the simplest physical systems that obey deterministic laws can admit unpredictable uncertainty (Robinson 1998; Batty and Longley 1994). Recent studies on nonlinear systems reinforced the effects of bifurcation and chaos, but they also suggested that orders exist within chaos (Batty and Longley 1994). The work carried out at the Brussels School produced important results about physical and chemical systems that were also applied to urban modelling (Prigogine and Stengers 1984). The main idea of the Brussels School was that, for an open system where there is an instant exchange of material and energy with its environment, the flow of energy or material could take the system to a state "far from equilibrium" (Prigogine and Stengers 1984: 141). The nonlinear process of interaction between different elements of the system can generate possibilities of bifurcation that may upset the global state of the system and lead to "order from fluctuations" (Prigogine and Stengers 1984: 178). It is the same nonlinear process that is responsible for the destruction of the order and the generation of a new order beyond another bifurcation. These ideas imply the existence of a self-organising capability in a complex system; they also suggest that "insignificant local behaviour can lead eventually to qualitatively different global structure" (Wu 1998a: 732).

Based on the understanding of the open system theory, the process of urban development is being looked at in new ways. A city can be viewed as an open and complex self-organising system that is far from being in equilibrium, and it exists in a constant exchange of goods and energy with other cities and its hinterland. The structure of this system emerges from local actions where uncoordinated local decision making may give rise to coordinated global patterns. Urban development is thus a spatially dynamic process, exhibiting some fundamental features of a self-organising system. Like the spread of a disease or fire, the edge of an urban area is constantly expanding toward the adjacent rural land. Previous urban form has an impact on present urban form, and it, in turn, will influence future urban patterns. This understanding suggests that a ground-up approach under the self-organising paradigm to address the local behaviour of the system is more realistic in modelling urban development, which has resulted in the emergence of a new class of simulation models (Benenson and Torrens 2004; Wu 1998a; Batty 1997, 1995; White and Engelen 1994, 1993; Couclelis 1985). These models include diffusion-limited aggregation (DLA), geo-simulation based on automata, and the agent-based models.

Diffusion-limited aggregation (DLA) is one of the most important models of fractal growth. It is based on principles of fractal geometry illustrating the irregular structure of a system that has the same degree of irregularity on all scales. This characteristic of irregularity is called self-similarity. The growth dynamics of a DLA model is through accretion at the periphery, which is remarkably simple. For example, in a plane space with only one immobile seed, a particle is launched from a random position far away from the seed, and this particle is allowed to diffuse on the space. If the particle touches the seed, it is instantly immobilised and becomes part of the aggregate. Then, similar particles are launched one by one, and each of them stops upon hitting the aggregate cluster. With more particles clumping together, the probability of new particles sticking in the neighbourhood of the cluster increases. Therefore, after launching a large number of particles, a cluster with a dendritic structure extending from the seed results. This model, first developed by two physicists, T. A. Witten and L. M. Sander (1981), illustrates a general class of behaviour that underlies many phenomena characterised by dendritic growth, such as growth of frost on a windowpane, lighting, and sparks. It was first introduced by Batty (1991) into urban growth modelling. Following the rules proposed by Witten and Sander (1981), Batty developed a model of DLA to model the dynamic urban growth, which was applied and tested in both small- to medium-sized cities (Batty and Longley 1994).

Urban models based on the automata technique have also emerged under the paradigm of a self-organising system, with cellular automata being the simplest but most popular in action. An automaton is an entity that has its own spatial and non-spatial characteristics but also has the mechanism for processing information based on its own characteristics, rules, and external input (Benenson and Torrens 2004). Cellular automata are a special type of automata that are individual automata, arranged in regularly tessellated space, for example, a regular grid. Information can be processed and transmitted between cells (or automata), which propagates through neighbouring automata. Although the technique of cellular automata dates back to the very beginning of digital computation (Macrae 1992), it is only since the late 1980s that this technique has been used to explore the behaviour of a self-organising system and to model the process of urban growth (Wu and Webster 2000, 1998; Batty, Xie, and Sun 1999; Batty 1998; Clarke and Gaydos 1998; Wu 1998a,b,c, 1996; Batty, Couclelis, and Eichen 1997; Batty and Xie 1997, 1994c; Clarke, Hoppen, and Gaydos 1997; Couclelis 1997, 1989, 1985; White and Engelen 1997, 1994, 1993; White, Engelen, and Uljee 1997; Cecchini 1996; Itami 1994).

Another type of automata, the multiple agent systems (MAS), has also been adopted for use in urban modelling (Torrens and Benenson 2005; Benenson, Omer, and Hatna 2002; Benenson 1999, 1998). The multi-agent systems are designed as a collection of interacting autonomous agents, each having its own capacities and goals, but together they relate to a common environment. This type of model operates on the same principles as the cellular automata model, with each agent being considered as individual autonomous agent-automata (Torrens 2003), and their states generally represent some agent-based characteristics. However, distinctions between cellular automata and multi-agent systems exist in a number of ways. One distinction is that in the multi-agent system, the basic unit of activity is the collection of agents representing individuals, developers, planners, or government decision-makers. The agents are autonomous in that they are capable of making independent actions, their activities are directed toward achieving defined tasks or goals, and their influence on the environment can be at different scales.

Another distinction between the cellular automata and the multi-agent systems is that cellular automata are fixed cells in the CA lattice, whereas the agents in the multi-agent systems are dynamic and mobile entities that can move within the spaces that they "inhabit" (Torrens 2003). These agents also can process and transmit information while they move along the spaces and pass the information from one agent and environment to another in their neighbourhood. Consequently, the neighbourhood relationships in agent automata are also dynamic: when individual agents alter their locations in space, their neighbourhood relationships also change. This modelling technique offers more flexibility as their agents not only carry the internal feature of the automata and the mechanism of transmitting information to its neighbours, they also represent the behavioural characteristics or can even simulate "intelligence" (Torrens 2003). Therefore, this type of modelling approach also offers attractive features in urban modelling (White and Engelen 2000). However, in practice, the agent-based models are less popular compared to CA-based models in urban simulation, and most of the agent-based models developed in urban modelling were actually formulated as CA and reinterpreted as multi-agent systems (Benenson and Torrens 2004; Torrens 2003; Benenson, Omer, and Hatna 2002).

1.3.2 Fuzzy Set and Fuzzy Logic

Another development in urban modelling in recent years is the application of the fuzzy set theory and fuzzy logic. With the understanding that precise mathematics are not always sufficient to model adequately the behaviour of complex systems, it was thought that a different kind of mathematics—the mathematics of fuzzy or cloudy quantities that were not describable in terms of probability distributions—was needed (Zadeh 1962). The fuzzy set theory was developed to handle problems that have no sharp boundaries or situations in which events are fuzzily defined. Fuzzy logic was used to describe the fuzzy relationships in a fuzzy system (Zadeh 1965). Fuzzy set and fuzzy logic were originally developed by Zadeh (1971, 1965, 1962), and they were applied in control engineering (Holmblad and Osterguard 1982, 1981; Mamdani and Assilian 1975). However, the potential applications of this theory go far beyond the design of intelligent controllers, and it is realised to be applicable to many data analysis, decision-making, and modelling practices relevant to geography (Openshaw and Openshaw 1997).

Urban development is the result of a series of decision-making processes that are featured by a range of uncertainties in these processes. Fuzzy sets and fuzzy logic provide ways of dealing with these uncertainties (Zimmermann 1987, 1985; Zadeh 1965). For a system represented by a series of "linguistic variables" (Wu 1998b, 1996; Altman 1994; Wang 1994; Zadeh 1975a,b,c), such as "good," "very good," or "moderate," a concept of membership degree is defined in a fuzzy set and is used to represent the extent of the uncertainties. Through the processes of "fuzzification" and "defuzzification," the behaviour of decision making can be modelled in a manner that better represents human thinking (Zadeh 1965). Applications of fuzzy set and fuzzy logic have been observed covering land suitability analysis (Davidson, Theocharopoulos, and Bloksma 1994; Hall, Wang, and Subaryono 1992). There is an increase in the literature looking at integrating fuzzy reasoning and fuzzy query into GIS (Wang and Hall 1996; Altman 1994; Wang 1994; Banai 1993; Kollias and Viliotis 1991). There is further literature regarding the application of fuzzy logic control to build dynamic simulation models (Wu 1998b, 1996).

1.3.3 GIS and Urban Modelling

Although GIS were developed over three decades ago and have been recognised as effective tools in geographical research since, these techniques had been developed in parallel to urban modelling without much interactions for over two decades (Sui 1998). It was not until the late 1980s that GIS researchers tried to integrate their techniques with urban modelling in the hope of improving the analytical capabilities of GIS techniques (Fischer, Scholton, and Unwin 1996; Fotheringham and Rogerson 1994; Anselin and Getis 1992; Fischer and Nijkamp 1992; Goodchild, Haining, and Wise 1992). Following these efforts, during the 1990s, both GIS users and urban modellers showed an increasing interest in the integration of the two techniques. Through this integration, urban modellers have recognised that GIS has provided modellers with new platforms for data management and visualisation (Nyerges 1995).

Many strategies for linking models with a GIS exist, which can be classified in a broad scale as a loose coupling or a strong coupling strategy. The loose coupling strategy is usually based on importing or exporting common data that are used in both the model and the GIS. On the other hand, a strong coupling strategy is based on adding the functionality of one system to the other either by embedding a model within a GIS or vice versa (Batty and Xie 1994a).

Other classifications on the integration of GIS and urban modelling approaches also exist. For instance, Sui (1998) identified four different approaches that have been widely used by researchers. These include embedding GIS-like functionalities into urban modelling packages, as those illustrated by Birkin et al. (1996), Putnam (1992), Clarke (1990), and Haslett, Wills, and Unwin (1990); embedding urban modelling into a GIS by software venders, such as those in the packages of TransCAD (Caliper Corporation 1983) and the ArcGIS Spatial Analyst Extension (ESRI 2004a); loose coupling an urban model with a GIS package where there are constant data exchanges between the two systems, such as those used by Clarke and Gaydos (1998); and the tight coupling of GIS and urban modelling via either a GIS macro or conventional programming, such as those developed by Batty and Xie (1994a,b), Anselin, Dodson, and Hudak (1993), Ding and Fotheringham (1992), and Miller (1991). Although GIS software vendors have increasingly recognised the importance of GIS's analytical and modelling capabilities, most of the GIS-based urban modelling efforts are made through the loose or tight coupling approach (Clarke and Gaydos 1998; Sui 1998).

With the shift of urban modelling from the conventional top-down approach to the current practices in addressing localities, models such as those using diffusion-limited aggregation and cellular automata techniques have demonstrated considerable potential in the mutual benefits of urban modelling and GIS (Batty, Xie, and Sun 1999; Wu 1998a,b, 1996).

1.4 PROBLEMS AND PROSPECTS

Although there has been a long history in developing mathematical models of urban development and progress has been made in current practice, the efforts of urban modelling have encountered several difficulties both theoretically and technically.

1.4.1 THEORETICAL PROBLEMS

From the review of the theoretical approaches to urban modelling in Section 1.2, it is obvious that the theoretical base of urban modelling is very weak. For instance, the urban ecological approach was based on the belief that human behaviour was determined by ecological principles, whereas the social physical approach was developed as a direct analogy to theories in physics. The neoclassical approach was based on the utility maximisation law in economics that was extended to the behavioural approach exploring the motivation of individual behaviour and decision-making processes. Each approach emphasised one or two aspects of urban development and was limited in describing other aspects. To what extent these theories are appropriate in explaining the patterns and processes of urban development remains a question. Although the systems approach has the advantage of exploring conceptually the complex relationships among factors within the urban system, this approach is unable to fundamentally explain the structuring and development of urban systems.

On the other hand, traditional modelling practices have been focused on the process of model design and the technical structure of the model. These modelling practices have contributed little to the development of new theories (Sui 1998; Batty 1981; Echenique 1975; Lee 1973). Recent studies in fractals and the Chaos Theory have led to the development of the cellular automata modelling (Batty 2000; Couclelis 1997, 1989, 1985; Toffoli and Margolus 1987). This approach has been widely applied as a modelling technique in physics, chemistry, biology, computer science, geography, and other environmental sciences (for instance, Clarke, Brass, and Riggan 1995; Mainster 1992; Doolen and Montgomery 1987; Guan 1987; Sander 1986; Vincent 1986; Burks and Farmer 1984; Hillis 1984; Vichniac 1984). However, both the theoretical base and the practical applicability of this approach need to be further developed and tested (Sui 1998).

Another theoretical problem in urban modelling concerns the behaviour and the sensitivity of models. As was discussed in Section 1.1.5 regarding the pitfalls of model building, it is most important to verify the model's outcomes and check if the model truly represents the real-world system it models. It is also necessary to test the sensitivity of the model, which usually includes tests of the model's output with different inputs, or the effects of errors in computation because of the rounding-up of values, or the effect of compounding cumulative errors when submodels are linked (Echenique 1975). This procedure of model verification and testing is especially important if the model is built on assumptions or hypotheses. Even if a model is developed based on sound theories, the model still needs to be tested rigorously as errors in the model may be larger than the expected change of the system being modelled (Echenique 1975). Unfortunately, most conventional urban models have not included these critical verification and validation tests. As a consequence, they could not be used as synthetic devices for testing urban policies or comparing growth scenarios in different cities.

In addition to the weak theoretical base and the lack of validity/sensitivity analyses in urban modelling, many models are developed based on a specific locality; they lack the general reapplicability to other regions. Moreover, most conventional models were developed to model cities in more developed countries. They lack the ability to explain the patterns and processes of urban development in the less developed countries, especially in the socialist countries. For example, the urban ecological approach was based on urban factorial ecology that pertained to Western societies. It is hard to define urban growth patterns in terms of a social area model in cities of socialist countries. Although urban development in less developed countries is very rapid in the recent years, there is an obvious lack in building models to describe the structure and patterns of urban growth in these countries compared to the efforts in modelling urban development in the more developed countries.

1.4.2 TECHNICAL PROBLEMS

Another type of problem with urban modelling concerns technical issues. One of these relates to data availability and data handling; the other relates to the way GIS can be applied in implementing urban models and in manipulating and visualising data.

Conventionally, data are collected by researchers undertaking specific research projects. With the applications of GIS and remote sensing technology, more and more data become available from both commercial data providers, government organisations, and professional institutions. However, problems exist in the comparability and accuracy of data and the way in which they are handled. These problems relate to the variation in the spatial areal unit that is frequently designed for different purposes, the level or degree of aggregation of spatial data, and the way of sampling for spatial data acquisition. The accuracy of data also depends on data providers and the way data are stored and presented. It is from this perspective that there exists a general lack of *good* data for specific research purposes. Therefore, searching for and achieving such *good* data in all modelling efforts becomes an important task.

The rapid growth of GIS and its integration with urban modelling has provided modellers with new platforms for data management and visualisation (Nyerges 1995). However, this integration is essentially technical in nature, and it has not touched upon fundamental issues in either urban modelling or GIS (Sui 1998). This is due to the difference in the spatial data representation schemes involved in urban modelling and GIS (Abel, Kilby, and Davis 1994). Essentially, the development of GIS is based upon a limited-map metaphor (Burrough and Frank 1995; Harris and Batty 1993), where geographical features in space are captured in map layers either as points, lines, and polygons, or as raster cells, and these features are temporally fixed (Raper and Livingstone 1995; Gazelton, Leahy, and Williamson 1992; Peuquet 1988). This scheme of representation is not compatible with the relative/relational and dynamic conceptualisation of space in urban modelling (Sui 1998). Therefore, the current effort in integrating a GIS with urban modelling is not satisfactory, and requires research at a higher level in conceptualising space and time, which has led to the emergence of the new geographical information sciences (Goodchild 1992; NCGIA 1995; Sui 1998).

1.4.3 FUTURE PROSPECTS

With the awareness of issues concerning urban modelling, improvements can be achieved in undertaking this practice. Theoretically, the development of self-organising systems and complexity theories has provided new ways of understanding the patterns and processes of urban development. A large body of literature has been established for constructing urban models under the new self-organising paradigm that has been applied and tested under various circumstances and in different regions. Through these efforts, the theoretical foundations of urban development will be understood through locally defined urban transitions.

The theoretical development in urban modelling will also benefit researchers in data collection and handling. As March (1974: 12) states, "we collect data intentionally, and behind these intentions are theories about how the world is constructed."

The establishment of certain models demands certain kinds of information (Echenique 1975). Therefore, researchers can focus on looking only for the information they need.

For the integration of urban modelling with GIS, Sui (1998) argued that the problems could not be solved if this integration continued to be treated as a technical issue. He suggested a new framework for urban modelling based on the newly developing geographical information science. Models under this framework should enable researchers to describe the emerging urban form in more comprehensive ways, to explain the underlying processes contributing to the emergence of new forms, and to prescribe effective urban policies to redirect the underlying process to promote the most desirable urban forms (Sui 1998).

1.5 CONCLUSION

This chapter provides an overview of the context on urban modelling and a theoretical as well as practical review of modelling techniques in urban development research. Continuing from this overview, the following chapters focus more on urban modelling based on the cellular automata approach. The structure of the book is as follows:

Chapter 2 first introduces the cellular automata approach and its application in urban modelling. Research on urban development based on the cellular automata is surveyed, and the problems raised by using this approach are identified. Based on the understanding of urban modelling and the applications of the cellular automata in this field, Chapter 3 introduces fuzzy set and fuzzy logic approaches in urban modelling, and it develops a cellular automata model of urban development using fuzzy constrained transition rules. A complete process of developing a fuzzy logic controller, from the fuzzification of input data to transition rule setting and fuzzy inferencing to the defuzzification of results, is presented in this chapter.

Chapters 4, 5, and 6 apply and calibrate the fuzzy constrained cellular automata model to simulate the process of urban development in the metropolitan area of Sydney, Australia. In Chapter 4, the descriptions of Metropolitan Sydney in relation to urban development and urban planning are addressed, followed by the construction of a database for the application of the fuzzy constrained cellular automata model of urban development. The urban development of Sydney during 1976–2006 is visualised in a GIS. By using the Sydney database, the cellular automata model of urban development is configured and calibrated in Chapter 5. Through vigorous testing and calibration, the model is used to simulate Sydney's urban development in a cellular environment and to evaluate the impacts of various factors on Sydney's urban development. The results generated by the model are presented and discussed in Chapter 6. Chapter 6 also presents discussions on the effects of spatial scales on the model's performances and outcomes. In addition, options for the future urban development of Sydney under various conditions from 2006 to 2031 are also simulated and presented in this chapter.

Finally, conclusions are drawn in Chapter 7 on urban modelling using the cellular automata approach and the application of the model to simulate the actual process of urban development. With these concluding remarks, future research directions are mapped out, thus bringing the book to a closure.

2 Cellular Automata and Its Application in Urban Modelling

Recent studies of nonlinear and open systems have led to the understanding of cities as evolutionary and complex systems (Allen 1997). Cities are looked at as self-organising systems, which are remarkably suited to computational simulation (Clarke and Gaydos 1998; Wolfram 1984). A cellular automaton is characterised by phase transitions that can generate complex patterns through simple transition rules. As such, this technique seems ideally suited to modelling the complexity of urban systems (Clarke and Gaydos 1998; Batty 1995). In this chapter, the principles of cellular automata simulation are discussed, and the applications of this simulation technique in modelling urban development are reviewed. Through this review, the progress and limitations of the cellular automata for urban development modelling are identified, leading to the development of a fuzzy constrained cellular automata model of urban development in the following chapters.

2.1 CELLULAR AUTOMATA MODELLING

2.1.1 CELLULAR AUTOMATA MODELLING: A GAME

A cellular automaton (CA) is a discrete dynamic system in which space is divided into regular spatial cells, and time progresses in discrete steps. Each cell in the system has one of a finite number of states. The state of each cell is updated according to local rules, that is, the state of a cell at a given time depends on its own state and the states of its neighbours at the previous time step (Wolfram 1984).

The research on the design and application of cellular automata dates back to the dawn of digital computation. Alan Turing, an English mathematician, first demonstrated that computers, through their software, could embody rules that could "reproduce" themselves (Batty 1997: 267). Stanislaw Ulam, a Polish-born American mathematician, studied the growth of crystals in the 1940s using a simple lattice network. At the same time, John von Neumann, Ulam's colleague at Los Alamos National Laboratory, was working on the problem of self-replicating systems. While working on his design, he realised the great difficulty of building a self-replicating robot and the high costs in providing the robot with a "sea of parts" from which to build its replicate. Ulam suggested that simple cellular automata could be found in sets of local rules that generated mathematical patterns in two-dimensional and three-dimensional space where global order could be produced from local action (Batty, Couclelis, and Eichen 1997; Ulam 1976). Inspired by Ulam, von Neumann constructed a complex self-producing machine with a two-dimensional cell space, each cell having 29 states that work within a small four-cell neighbourhood, that is, the East/South/West/North adjacent cells. This four-cell neighbourhood is therefore called the *von Neumann Neighbourhood*. von Neumann proved that, mathematically, with his machine, any particular pattern or blueprint would make endless copies of itself within the given cellular space (von Neumann 1966). von Neumann's work set alight the field in the 1950s, initiating the scientific study of cellular automata (Batty 1997). However, owing to its complexity, von Neumann's machine was not run under a "real" simulation on any modern digital computer (Langton 1984).

The first important application of the cellular automata came from John Conway's "Game of Life" (Gardner 1972). "Life" was constructed as a two-dimensional grid with two cell states and an eight-cell neighbourhood. The two possible states a cell can be in is either dead or live. The eight-cell neighbourhood includes cells in the East, South, West, North, South-west, South-east, North-east, and North-west directions. This type of neighbourhood is termed the *Moore Neighbourhood*.

In Conway's "Game of Life," a cell can survive, die, or give birth in successive generations according to the following rules:

- Survival: A live cell with two or three live neighbours survives into the next generation.
- Death: A live cell with less than two or more than three live neighbours dies either of isolation or of overcrowding.
- Birth: A dead cell with exactly three live neighbours becomes alive in the next generation.

Using these simple rules, the model is able to generate very complex structures as different cells die, survive, or give birth in successive generations. Figure 2.1 presents a sample of simulation results generated by the model. The "Game of Life" has been a very popular cellular automata model after the paper by Gardner in *Scientific American* (Gardner 1972).

Conway's "Game of Life" has drawn great interest from a whole generation of researchers. Countless approaches have been developed to explore the complexity of cellular automata that emerges from simple rules. In 1983, Stephen Wolfram published the first of a series of papers systematically investigating the simplest one-dimensional cellular automata, which he termed *elementary cellular automata* (Wolfram 2002, 1994, 1983). Each cell in the elementary cellular automata only has two possible values or states of either 0 or 1, and the transition rules depend only on the nearest-neighbour values. The unexpected complexity of the behaviour of these simple rules led Wolfram to suspect that complexity in nature may be due to similar mechanisms (Wolfram 1994).

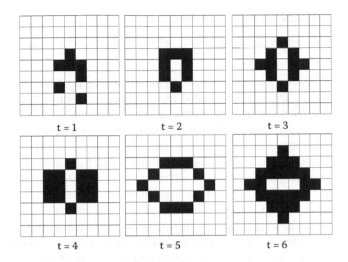

FIGURE 2.1 A simple simulation based on Conway's "Game of Life." (Black cells are live, and white cells are dead; *t* is time step.)

However, studies on cellular automata had not taken off until the 1990s when computers had become truly graphic and studies on complexity theory, self-organisation, and chaos had reached fever pitch (Batty 1997). From the viewpoint of complexity theory, complex structures can be generated by very simple rules; therefore, they can provide useful techniques for exploring a wide range of fundamental issues in dynamics and evolution. As Wolfram shows, cellular automata are constructed "from many identical components, each simple, but together capable of complex behaviour" (Wolfram 1984: 419). Cellular automata have been applied primarily in the physical and natural sciences such as physics, chemistry, and biology, and there is also an increasing number of applications to cities and spatial ecological studies (Batty 1997; Itami 1994).

2.1.2 A Simple Cellular Automata Model

According to Levy (1992), a cellular automaton is a self-operating machine that "processes information, proceeding logically, inexorably performing its next action after applying data received from outside itself in light of instructions programmed within itself" (Levy 1992: 15). In a cellular automata system, space is divided into regular cells. The state of a cell is determined by the state of the cell itself and the states of cells in its neighbourhood at a previous time step through a set of locally defined transition rules. The states of all cells are updated synchronously. The overall behaviour of the system is determined by the combined effects of all the local transition rules. Thus, the state of the system advances in discrete time steps.

2.1.2.1   Five Basic Elements of a Cellular Automaton

According to the foregoing definition, a cellular automaton consists of five basic elements:

a. *The cell,* which is the basic spatial unit in a cellular space. Cells in a cellular automaton are arranged in a spatial tessellation. A two-dimensional grid of cells is the most common form of a cellular automaton used in modelling urban growth and land-use change. However, other arrangements, such as a one-dimensional cellular automaton, have also been developed to represent linear objects, such as urban traffic modelling (Nagel, Rasmussen, and Barrett 1997; DiGregorio et al. 1996). The cell space can also be tessellated into other arrangements, such as a honeycomb arrangement or even in three dimensions. For instance, the third dimension of an urban cellular automaton can represent the building heights in the urban built-up environment. However, due to the difficulty in model design and construction, those cell arrangements are less popular or yet to be developed in urban modelling practice.

b. *The state,* which defines the attributes of the system. Each cell can take only one state from a set of states at any one time. The state can be a number that represents a property. In urban-based cellular automata models, the states of cells may represent the types of land use or land cover, such as urban or rural, or any specific type of land use; or it may be used to represent other features of the urban area, such as social categories of populations (immigrant vs. native population) as was proposed by Portugali and Benensen (1995).

c. *The neighbourhood,* which is a set of cells with which the cell in question interacts. In a two-dimensional space, there are two basic types of neighbourhoods: the von Neumann Neighbourhood (four cells), which includes the North, South, East, and West neighbours of a cell in question; and the Moore Neighbourhood (eight cells), which includes the cells defined in the von Neumann neighbourhood as well as cells in the North-west, North-east, South-east, and South-west directions. Other kinds of neighbourhoods, such as a neighbourhood within a circle of a certain distance from the cell in question, have also been used in urban modelling (e.g., see White and Engelen 1997, 1994).

d. *The transition rule,* which defines how the state of one cell changes in response to its current state and the states of its neighbours. This is the key component of cellular automata because these rules represent the process of the system being modelled, and are thus essential to the success of a good modelling practice (White 1998). For a strict cellular automaton, the transition rules are uniform and are applied synchronously to all cells within the system. However, a number of modifications in defining transition rules have been observed in the literature, which will be discussed in the following sections.

e. *The time,* which specifies the temporal dimension in which a cellular automaton exists. According to the definition of cellular automata, the states of all cells are updated simultaneously at all iterations over time. However, this restriction can be released by operating the cellular automata model at different temporal speed for different cells, as was seen in Uljee, Engelen, and White (1996) where their model simulated the low-lying areas on a monthly basis and the upland areas on a yearly basis.

2.1.2.2   Mathematical Representation of a Cellular Automaton

Let $S_{x_{ij}}^t$ be the state of a cell x_{ij} at the location i, j at time t. $S_{x_{ij}}^t$ belongs to a finite number of states of cells in the cellular space. Let $S_{x_{ij}}^{t+1}$ be the state of the cell at time $t + 1$. Then,

$$S_{x_{ij}}^{t+1} = f\left(S_{x_{ij}}^t, S_{\Omega_{x_{ij}}}^t \right) \tag{2.1}$$

where $\Omega_{x_{ij}}$ represents a set of cells in the neighbourhood of cell x_{ij}, $S_{\Omega_{x_{ij}}}^t$ is a set of states of cells $\Omega_{x_{ij}}$ at time t, and f is a function representing a set of transition rules.

If the cell itself is considered as a member of its neighbourhood, then Equation 2.1 can be rewritten as

$$S_{x_{ij}}^{t+1} = f\left(S_{\Omega_{x_{ij}}}^t \right) \tag{2.2}$$

Equation 2.2 can be expressed in a verbal form that illustrates a generic principle of the development of a cellular automaton, namely,

IF   something happens in the neighbourhood of a cell,
THEN   something else will happen to the cell at the following time step.

A cellular automata model usually consists of a set of "IF–THEN" statements that imply specific transition rules. For instance, the model "Game of Life" can be expressed as three "IF–THEN" statements:

IF   there are two or three live cells in the Moore Neighbourhood of
              a live cell,
THEN   the cell stays alive in the next generation;
IF   there are less than two or more than three live cells in the Moore
              Neighbourhood of a live cell,
THEN   the live cell dies in the next generation;
IF   there are exactly three live cells in the Moore Neighbourhood
              of a dead cell,
THEN   the dead cell becomes alive in the next generation.

Owing to the generic principle of development, cellular automata models "may serve as a framework for modelling complex natural phenomena in a way that is conceptually clearer, more accurate, and more complex than conventional mathematical systems" (Itami 1994: 30).

2.1.3   The Complex Features of Cellular Automata

One of the most attractive features of the cellular automata is its capability in generating complicated behaviour and complex global patterns despite its simplicity in model construction. For instance, even the simplest or "elementary" one-dimensional cellular automaton with only binary values at each site was capable of exhibiting complicated behaviour (Wolfram 1983). This indicates that simple models such as cellular automata can potentially be made to reproduce complex phenomena (Wolfram 1994).

Cellular automata also possess the characteristic of an open system that is capable of self-organisation. This is a process in which the system increases its internal organisation in complexity without being guided or managed by an outside source. Such a self-organising system typically displays emergent properties in which novel and coherent structures and patterns arise from turbulence and chaos (Goldstein 1999). For instance, Wolfram (1983) shows that with a "disordered" initial state that was randomly chosen, a simple one-dimensional cellular automaton is capable of generating some structure in the form of many triangular "clearings." The spontaneous appearance of these clearings is a simple example of "self-organisation" (Wolfram 1994).

Another generic feature of the cellular automata is its capability to generate self-similar patterns (Wolfram 1994). That is, as the cellular automata develop over time, the patterns they generate often exhibit a degree of regularity in structure, which are self-similar (Torrens 2000). With this feature, it indicates that portions of the evolved pattern of a structure are indistinguishable from the whole; therefore, the structure of the pattern is scale independent (Torrens 2000; Wolfram 1994), making it attractive for overcoming the modifiable area unit problem that exits in most geographical modelling practices (Openshaw 1984).

In addition, cellular automata are dynamic systems, and they attend to elements of the system they represent at a local scale. Through the interaction with other elements within their neighbourhood the cellular automata are capable of modelling complex phenomena by setting simple transition rules. As such, they can be used as simple mathematical idealisations of natural systems (Wolfram 1994).

2.2 CELLULAR AUTOMATA IN URBAN MODELLING

Urban development resembles the behaviour of a cellular automaton in many aspects. The space of an urban area can be regarded as a combination of a number of cells, each cell taking a finite set of possible states representing the extent of its urban development. The state of each cell evolves in discrete time steps according to some local rules. This section identifies the advantageous features of cellular automata in urban modelling as well as its early and contemporary applications in urban modelling practices.

2.2.1 An Urban Cellular Automata

Let us consider an imaginary city constructed in a cellular space. This city consists of a two-dimensional regular grid of n × n cells, or land parcels. Each land parcel may have one of two possible states: urban or non-urban. The neighbourhood represents a region that impacts on the development of the parcel in question. The transition rules determine how a land parcel transits from one state to another, hence implying the process of a development in the locale. These transition rules are usually expressed as a set of "IF–THEN" statements, that are intrinsically simple. However, these simple rules can generate complex patterns of development.

We first assume uniform social, economic, and environmental conditions throughout the whole region. That is, apart from a few land parcels that are urbanised (those parcels are displayed in black on Figure 2.2, $t = 0$), their state is presented

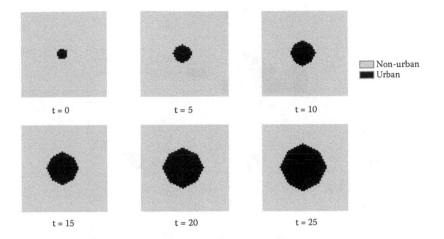

FIGURE 2.2 A cellular automata-generated urban development in a plain area. (Black cells are urban; grey cells are non-urban; *t* is time step.)

as urban; all the rest of the parcels are in a non-urban state, with identical socio-economic and environmental conditions. Therefore, the only factor that will drive the development of land parcels is the number of developed parcels in the neighbourhood of a parcel in question, implying the growth of any new urban parcels that might be influenced by the urbanised land parcels. Using the Moore Neighbourhood, the transition of the state of parcels is governed by the following rule.

Rule 1:

IF   there are three or more developed parcels (i.e., urban parcels) in the Moore Neighbourhood of a non-urban land parcel in question,

THEN   the non-urban land parcel will be developed into an urban state.

With this transition rule, the model generates a series of scenarios of urban development at different time frames, which are displayed in Figure 2.2.

However, in a real situation, the geographical conditions within an area can never be uniform. For instance, significant difference may exist in the terrain of the landscape. In order to reduce the costs in the construction and operation of municipal facilities such as sewage and water supply, urban development may be restricted to areas with a relief of less than 300 m. Therefore, no development will take place in cells with a relief of more than 300 m. In this situation, a new rule needs to be implemented in the model to reflect the terrain restriction. This new rule can be presented as another IF–THEN statement.

Rule 2:

IF   the relief of the landscape is more than 300 m,

THEN   the land parcel will remain undeveloped (i.e., it stays as a non-urban parcel).

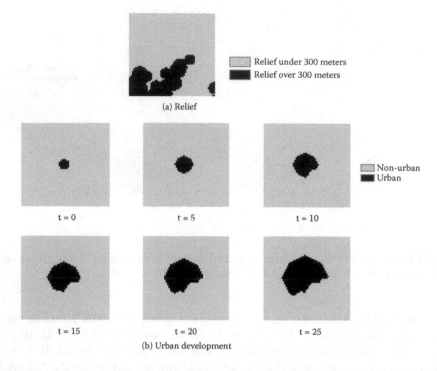

(a) Relief

(b) Urban development

FIGURE 2.3 Simulated urban development with terrain constraint. ((a) Relief (black: more than 300 m; grey: less than 300 m); (b) Urban development (black: urban parcels; grey: non-urban parcels; t: time step).)

With both Rules 1 and 2, the scenario of urban development in this area will change, as shown in Figure 2.3.

In addition, variation may also exist in the transportation network. For instance, if there is a major road running through the city, development might be attracted to areas along the road. In this case, another transition rule needs to be added into the model to reflect the effect of support by the transportation to urban development. This can be presented as the following IF–THEN statement:

Rule 3:
IF   there are one or two developed parcels (urban) in the Moore Neighbourhood of a non-urban parcel, and there is a road running through that parcel,
THEN   the non-urban parcel will be developed into an urban state.

Again, with the implementation of this new rule, the pattern of development in this area changes, as illustrated in Figure 2.4.

Using this modelling framework, more transition rules can be added into the model to reflect the various support or restrictions of social, economic, or environmental

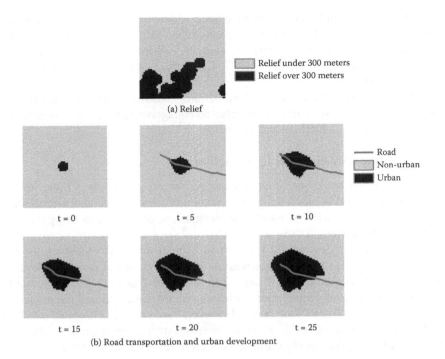

(a) Relief

Relief under 300 meters
Relief over 300 meters

t = 0   t = 5   t = 10

Road
Non-urban
Urban

t = 15   t = 20   t = 25

(b) Road transportation and urban development

FIGURE 2.4 Simulated urban development with transportation support. ((a) Relief (black: more than 300 m; grey: less than 300 m); (b) Road transportation (grey line); urban development (black: urban parcels; grey: non-urban parcels; *t*: time step).)

factors on urban development. The example here illustrates a very simplified model of urban development; however, it provides a general idea of how locally defined transition rules can be implemented in an urban cellular automata, and how simple rules can be applied to simulate the complex behaviour of an urban system in a cellular space.

2.2.2 ADVANTAGES OF CELLULAR AUTOMATA FOR URBAN MODELLING

With the full development of computer graphics, geographical information systems, fractals, and chaos and complex systems theories since the late 1980s, the applications of cellular automata to urban modelling are rapidly gaining favour among urban researchers. This is because cellular automata are intrinsically spatial, which is inherently attractive for their application to geographical problems (White and Engelen 1993). Cellular automata are "especially appropriate in urban modelling where the process of urban spread is entirely local in nature and the aggregate effects, such as growth booms, are *emergent*" (Clarke and Gaydos 1998: 700), that is, their behaviour is generated "by repetitive application of the rules beyond the initial condition" (Clarke and Gaydos 1998: 700). With these understandings, cellular automata have been widely used in geographical modelling, especially in studies of urban development (e.g., Clarke and Gaydos 1998; Clarke, Hoppen, and Gaydos 1997; Wagner 1997; White and Engelen 1997, 1994, 1993; White, Engelen, and Uljee 1997; Cecchini 1996; Itami 1988; Hillier and Hanson 1984).

Apart from the intrinsic nature of cellular automata toward urban modelling, this approach also possesses a number of other advantageous features that are attractive to urban modellers, which are outlined below.

2.2.2.1 Simplicity in Model Construction

Compared to all other urban simulation models discussed in Chapter 1, the development of urban models based on cellular automata are very straightforward, which can be constructed based on intuitive understanding of the system being modelled (Benenson and Torrens 2004).

As illustrated in Section 2.2.1, an urban cellular automata model can be constructed as a simple two-dimensional array of cells with one of two possible states: urban or rural. The transition of cells from one state (rural) to another (urban) is based on a number of simple rules, which can be implemented into the model as a set of simple "IF–THEN" statements. However, based on the self-organisation and self-reproduction natures of the cellular automata, such a simple design of the cellular automata can generate very complex spatial patterns when the system progresses over time.

The simplicity and intuitive nature of the cellular automata not only simplifies the process of model construction but it also makes it easier for modellers to understand the development of the system and interpret the model's results. This is because the model mimics the way in which "we study, understand and describe the system and phenomena in the real world" (Benenson and Torrens 2004: 11).

2.2.2.2 Modelling Spatial Dynamics to Support "What If" Experiments

Unlike most conventional urban models that focus more or less on the spatial patterns of urban growth, cellular-automata-based urban models usually pay more attention to simulating the dynamic process of urban development and defining the factors or rules driving the development. This is due to the characteristic that the cellular automata approach provides ways for dynamic modelling. It is particularly well suited to modelling complex dynamic systems composed of large numbers of individual elements. By applying different transition rules, a model based on cellular automata seeks to explore how the urban system has been developing and how this system changes under certain rules or forces. Therefore, it provides an environment to support "what if" experiments. This allows users to explore various possible futures and develop insights that may be of use in urban planning (White and Engelen 1997). For instance, White and Engelen (1994, 1993) and White, Engelen, and Uljee (1997) used cellular automata to simulate the process of the evolution of urban land-use patterns in the urban area of Cincinnati, the United States. Clarke, Hoppen, and Gaydos (1997) developed a cellular automata model to simulate the historical urban development process in the San Francisco

Bay area; they also applied the model to predict the long-term urban growth process in the Washington/Baltimore region in the United States (Clarke and Gaydos 1998). Wu (1998a,b,c, 1996) developed models based on cellular automata to simulate the dynamic process of land development in a fast-growing urban region in south-east China. These applications reveal that the foci of cellular automata models are the rules leading to the development of the system and the experimentation on how these rules affect the behaviour of the system. The spatial pattern or structure of the system is presented as the result of the dynamic spatial development process.

2.2.2.3 A "Natural Affinity" with Raster GIS

In modelling geographical phenomena and processes, most cellular automata models are constructed based on regular spatial tessellation; these models are naturally related to raster-based GIS. As shown by White and Engelen (1994) and Wagner (1997), advantages come with the integration of GIS and cellular automata models. Even using a loose coupling approach to integrating GIS and a cellular automata model, GIS can provide spatial data that form the initial configuration in a cellular automata model, and the simulation results can be returned to a GIS for further processing, visualisation, and storage. The application of cellular automata in modelling urban development is "virtually impossible without the data management capabilities of GIS" (Clarke and Gaydos 1998: 700). Therefore, applications of cellular automata in geography are mostly integrated with a GIS.

However, the integration of GIS with cellular automata models is more than just data exchange, storage, and visualisation. The similarities between cellular automata and raster GIS models suggest the strong potential for a complete integration of the two technologies. Wagner (1997) examined the advantages of integrating these two systems. He illustrated the possibilities of creating a limited cellular automata model within a GIS or implementing the analytical functionality of a GIS in a cellular automata system. By means of a cellular automata machine (CAM) as the analytical engine for GIS operations, Wagner developed a prototype to take the functionality of both GIS and a cellular automata machine. Takeyama and Couclelis (1997) developed a way of integrating cellular automata and a GIS through a geo-algebra, a mathematical generalisation of map algebra capable of expressing a variety of dynamic spatial models and spatial data manipulations within a common framework. Batty, Xie, and Sun (1999) also developed a software program to implement a GIS-based cellular automata model to simulate urban dynamics. These literatures suggest ways of a strong coupling of the two technologies, which are still under investigation.

2.2.3 EARLY APPLICATIONS OF CELLULAR AUTOMATA IN URBAN MODELLING

The application of cellular automata to urban systems can be traced back to the beginning of the cellular automata themselves — to the first attempt to build mathematical models of urban systems in the 1950s (Batty, Couclelis, and Eichen 1997). At that time, although much attention had been paid to the construction of models of urban spatial patterns under the social physical approach, some researchers built models to understand the process responsible for the formation of urban spatial patterns.

Torsten Hägerstrand's innovation-diffusion models were of this trend (Hägerstrand 1952). From these, via models of migration and of local settlement networks (Morrill 1965), the spatial growth of urban areas and the changes in urban structure were treated as different types of diffusion processes. In the construction of the diffusion models, a notion of the neighbourhood effect was included (Hägerstrand 1967, 1965); this was very close in principle to the cellular automata technology.

In the early 1960s, with the assistance of computer technology, models of urban growth were developed under the behavioural approach with more emphasis on individuals' behaviour and their decision-making processes. One of these models presented by Lathrop and Hamburg (1965) was developed in a cell-based frame-work to simulate the development of an urban area in western New York State. This model was relevant in spirit to the principles of change in a cellular space. Another model, developed by Chapin and his colleagues at the University of North Carolina in modelling the process of land development, articulated cell-space modules where changes in states were predicted as a function of a variety of factors affecting each cell, some of which embodied neighbourhood effects (Chapin and Weiss 1968).

The application of cellular automata in urban models came from theoretical quantitative geography. Waldo Tobler was the first scholar to propose a cell-space model simulating urban growth in the Detroit region (Tobler 1970). Using his first rule of geography that "everything is related to everything else, but near things are more related than distant things" (Tobler 1970: 236), this model attempted to relate the population growth of a cell (representing an area of one-degree quadrilaterals of latitude and longitude) to the population of the same and neighbouring cells in the immediately preceding time period. Following this research, Tobler began to explore the way that general cellular automata could be used in geographic systems. In his famous paper "Cellular Geography," Tobler (1979) summarised the five types of models of land-use change and systematically described the notion of cellular-based geographical modelling (Figure 2.5).

Within Tobler's five types of models, the geographical model (Model V) shows that the land use at location i, j is dependent on the land use of the model itself and all the land uses in the neighbourhood of the location i, j.

$$g_{ij}^{t+\Delta t} = F\left(g_{i\pm p,j\pm q}^{t}\right) \tag{2.3}$$

where g_{ij}^{t} is the land-use category (urban, rural, ...) at location i, j at time t; $g_{ij}^{t+\Delta t}$ is the land-use category at the same location at some other time; $g_{i\pm p,j\pm q}^{t}$ represents all the land-use categories in the neighbourhood of the location i, j. This model established a theoretical framework for the application of cellular automata in geography, and urban development modelling in particular.

Influenced by Tobler, Couclelis (1985) developed a model with a geographical interpretation for Conway's "Game of Life." In that model, Couclelis explained that a "live" cell could be an urban zone with a threshold population needed to support a school. A zone could retain its school population base if two or three of the neighbouring zones also exceeded the threshold (survival); it would lose its school if four or more of the neighbouring zones had an above-threshold population, or if the population declined drastically in that whole neighbourhood, suggesting abandonment of the

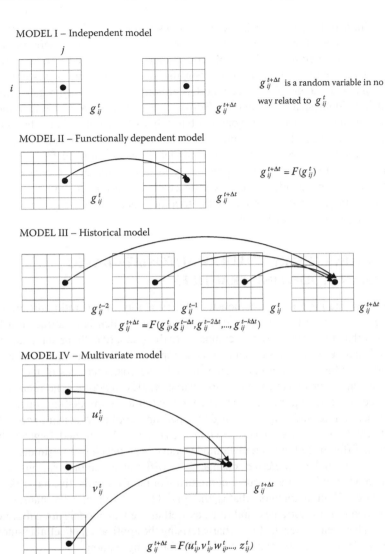

MODEL I – Independent model

$g_{ij}^{t+\Delta t}$ is a random variable in no way related to g_{ij}^t

MODEL II – Functionally dependent model

$$g_{ij}^{t+\Delta t} = F(g_{ij}^t)$$

MODEL III – Historical model

$$g_{ij}^{t+\Delta t} = F(g_{ij}^t, g_{ij}^{t-\Delta t}, g_{ij}^{t-2\Delta t}, \dots, g_{ij}^{t-k\Delta t})$$

MODEL IV – Multivariate model

$$g_{ij}^{t+\Delta t} = F(u_{ij}^t, v_{ij}^t, w_{ij}^t, \dots, z_{ij}^t)$$

MODEL V – Geographical model

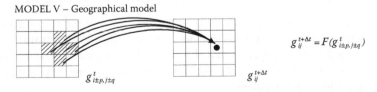

$$g_{ij}^{t+\Delta t} = F(g_{i\pm p, j\pm q}^t)$$

FIGURE 2.5 Graphic illustration of Tobler's five models using a 25-cell geographical array. (Adapted from Tobler, W. R., Cellular geography, In *Philosophy in Geography,* Ed. S. Gale and G. Olsson. D. Reidel, Dordrecht, The Netherlands, 379–86, 1979. With permission.)

area (death); and three dense neighbouring zones would provide for a zone to reach the necessary population base and thus acquire a school (birth) (Couclelis 1985).

Couclelis also explored the limits of cellular models in geography and generalised theoretically a cellular automata formalism that seemed capable of systematically integrating a wide range of advanced models at both the level of aggregate phenomena and individual spatial behaviour. However, her model was not understood as realistic statements of urban development but only as "metaphors" of urban growth (Couclelis 1997: 165). It was not until the 1990s that Couclelis identified how cellular automata-based urban and regional models could be moved from the realm of instructive metaphors to that of potentially useful quantitative forecasting tools through generalising the idea of space within a cellular automaton to proximal space, and the operations of a cellular automaton to a more general geo-algebra (Couclelis 1997; Takeyama and Couclelis 1997).

2.3 CONTEMPORARY CELLULAR AUTOMATA-BASED URBAN MODELLING PRACTICES

Cellular automata models have demonstrated their ability in generating different spatial patterns based on locally defined transition rules, which have attracted a lot of interest in urban research. However, compared to the standard cellular automata model that strictly defines its five basic elements, contemporary cellular automata-based modelling practices have broadly interpreted the defining characteristics of the standard cellular automata, or relaxed some of its characteristics to address the requirements of a particular modelling problem. This new school of urban modelling started in the 1990s, when White and Engelen first developed and applied a constrained cellular automata model to simulate land-use change dynamics (White and Engelen 1993). Since then, different cellular automata models have been developed to simulate urban growth and urban land-use/cover change. The differences among various cellular automata models exist in their way of configuring or modifying the five basic elements of a standard cellular automaton, that is, the spatial tessellation of cells, states of cells, neighbourhood, transition rules, and time, as well as the way models are calibrated or validated. This created new features characterising the application of cellular automata to urban modelling, which are discussed in the following section.

2.3.1 SPACE TESSELLATION: FROM REGULAR TO IRREGULAR SPATIAL UNITS

According to the strict rules of cellular automata, a cellular space is tessellated into an array of regular cells; the spatial resolutions of the cells vary from small to very large cell sizes. More recently, research also shows that irregular spatial units can be applied to model urban development, which may yield better understandings of the urban system and its development.

2.3.1.1 Regular Cells of Small or Large Resolution

For models of spatial phenomena, scenarios resulting from the tessellation of space at different scales vary (Figure 2.6). This is commonly regarded as a modifiable area unit problem (MAUP). Openshaw (1984) provides a comprehensive review

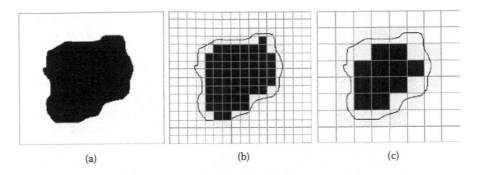

(a)   (b)   (c)

FIGURE 2.6 The modifiable area unit problem affecting spatial modelling. (Black pixels are urban; grey pixels are the boundary between urban and non-urban; white pixels are non-urban. (a) An urban area sited in a regional context; (b) the area tessellated into cells at a small scale; (c) the area tessellated into cells at a large scale. This figure shows that different patterns of urban, fringe, or non-urban can be achieved by tessellating the urban area at different scales.)

of the early research into the modifiable area unit problem. He pointed out that the MAUP is a potential source of error that can affect spatial studies that utilise aggregate data sources. Fotheringham and Wong (1991) explain that the modifiable area unit problem is essentially unpredictable in its intensity and effects in multivariate statistical analysis and is therefore a much greater problem than in univariant or bivariant analysis. Because cellular automata models are based on the tessellation of space into regular (or irregular) cells, a fundamental question is at what scale can the space be subdivided into spatial units to build up the model? Practically, the configuration of the spatial cell scale in urban modelling based on cellular automata is pragmatic: no specific consideration has been given to the tessellation of cell scale.

In the early applications of cellular automata in urban modelling, the resolution of cells seemed to be less an issue of concern. The use of cell resolution was either based on the availability of data or on the convenience for computation. Therefore, the resolution of cells varies significantly from one application to another. For instance, White and Engelen (1993) used a 500-m cell resolution to test a constrained cellular automata model to simulate the urban land-use patterns in four U.S. cities, including Atlanta, Cincinnati, Houston, and Milwaukee. Later on, they modified the model to a "high-resolution scale" with a 250-m grid resolution to simulate urban land-use dynamics in the city of Cincinnati, Ohio (White and Engelen 2000; White and Engelen 1997; White, Engelen, and Uljee 1997).

In other research, Wu (1996) used a 28.5-m cell resolution based on the Landsat imagery to simulate the urban land-use change in China's Guanzhou City. He also applied a 200-m cell scale to build a cellular automata model with transition rules derived from a multicriteria evaluation model to simulate the urban development within the same region in China. However, the reason for choosing a coarse resolution of 200 m was to save computation time; there was no justification on the selection of the different cell scales, nor were the results generated by the models compared and discussed.

Another cellular automata model developed by Clarke and colleagues used a basic grid of 300-m cells to simulate the urban growth in the San Francisco Bay area. However, while applying the model to the Washington/Baltimore region, calibrations were undertaken at different cell scales, including 210 m, 420 m, 840 m, and 1680 m (Clarke and Gaydos 1998). Their results show that, although not all rules or factors are sensitive to the change of the cell scale, the scale of cells does have an impact on the results of the simulation, especially in relation to certain factors, such as road and slope. They suggested a hierarchical approach in calibrating the model by "first using coarse data to investigate the scaling nature of each parameter in a different city setting, then scaling up once the best data ranges are found" (Clarke and Gaydos 1998: 710). This finding was supported by Samat (2006) who identified that the cellular automata model produces realistic urban form only up to a certain range of spatial resolution, indicating the selection of cell scale to ensure that the output produced maintains the overall accuracy of the model and morphology of urban areas.

Furthermore, Dietzel and Clarke (2004) studied how the change of spatial resolution affects the parameter setting of the cellular automata model and the outcomes of the model. Their results show that, by calibrating the model using finer spatial resolution data, the model will generate a different set of parameter space, which may result in different model outcomes.

Other researchers, such as Menard and Marceau (2005) and Kocabas and Dragicevic (2006), studied the combined impact of the spatial scale of cells and neighbourhood size on the model's performance and outcome. This will be discussed in Section 2.3.3 under "Neighbourhood definitions."

2.3.1.2   Using Irregular Spatial Units

Although it is typical to use regular, two-dimensional and homogeneous grids in cellular automata-based models, researchers also suggested the use of irregular spatial units in urban modelling (Yeh and Li 2006; O'Sullivan 2001; White and Engelen 2000; Couclelis 1985). For instance, White and Engelen (2000) pointed out that "if the space to be modelled is already subdivided into functionally relevant units that approximate the scale of the grid, then these units will provide a better representation of the space than will grid cells" (White and Engelen 2000: 387). However, largely due to the complications in the definition of the irregular neighbourhood and the computationally intensive operation required, only a few literatures were found to have actually implemented irregular spatial units in their urban modelling practice. One of such examples is by Batty and Xie (1994c), who employed cadastral units to model land-use change in a suburban area of Buffalo, United States.

Another literature by Shi and Pang (2000) presented a Voronoi-based cellular automata model to simulate the dynamic interactions among spatial objects. Similar to a vector-based spatial model, the Voronoi spatial model was designed to represent spatial objects as vector objects. The model was considered capable of simulating the interactions among point, line, and polygons with irregular shapes and sizes in a dynamic system; it can also operate on any form of spatial unit, such as square or triangular pixels, or real objects such as the location of a fire station, street, or land parcels (Shi and Pang 2000). Compared to the standard cellular automata, the Voronoi-based cellular automata is considered to be "a more natural and efficient representation of human knowledge over space" (Shi and Pang 2000: 455). However, the spatial units or polygons generated through the Voronoi approach do not represent the actual spatial units such as cadastral land parcels, or urban administrative units such as postal code areas or census tracts (Stevens and Dragicevic 2007).

A recent study by Stevens and Dragicevic (2007) demonstrated a cellular automata model using an irregular spatial structure with high spatial and temporal resolutions to simulate urban land-use change. Featured as the iCity model, it uses high-resolution land-use data with the spatial units composing irregularly sized and shaped cadastral parcels in vector structure. The model resulted in a better conformation of simulated results to actual land-use boundaries. However, their model also lacked computational efficiency, resulting in long processing times when running the model even for a small area (Stevens and Dragicevic 2007).

2.3.2 FROM BINARY AND MULTIPLE TO CONTINUOUS CELL STATES

Most of the cellular automata models of urban growth configure the state of cells using land-use or land cover type. For instance, White and Engelen (2000, 1997, 1993) defined a hierarchy of five land-use states, ranging from vacant (the lowest state), housing, industry, to commerce (the highest state). A cell can only develop from a lower state to a higher state, indicating the growth of the urban areas.

Other researchers used binary states to simulate the process of non-urban to urban conversion. The state of non-urban or urban can be defined in terms of land use (Li and Yeh 2000; Clarke and Gaydos 1998; Clarke, Hoppen, and Gaydos 1997; Wu 1998b,c, 1996); it can also be defined based on other spatially distributed variables, such as population densities or other socio-economic indicators (Wu 1998a). A common feature of these models is that the cells in a cellular space are defined as discrete states; there exists a clear boundary between each of the states. However, in practice, this sharp boundary between cell states may be difficult to identify.

To overcome this problem, research also shows the use of a continuous cell state based on fuzzy set theory to simulate the spatial dynamics of urban growth (Liu and Phinn 2003). This will be discussed further in the following chapters.

2.3.3 NEIGHBOURHOOD DEFINITIONS

2.3.3.1 "Action-at-a-Distance" Neighbourhood

According to standard cellular automata rules, the state of a cell is determined by the state of the cell itself and the states of cells in its neighbourhood at a previous time step. This neighbourhood is defined as cells immediately adjacent to the cell in question, such as the von Neumann Neighbourhood and the Moore Neighbourhood. Every change in state must be local. Therefore, only cells adjacent to the central cell have an impact on the transition of the state of the central cell. In other words, there is no "action-at-a-distance" in a standard cellular automata model (Batty, Couclelis, and Eichen 1997: 160).

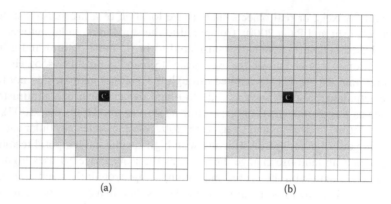

(a)   (b)

FIGURE 2.7 Large neighbourhood sizes used by White and Engelen (1993) and Wu (1996). ((a) A circular neighbourhood of 113 cells proposed by White and Engelen (1993); (b) a rectangular neighbourhood of 120 cells proposed by Wu (1996).)

Although most cellular automata models of urban growth still follow this strict rule when defining a neighbourhood size, some applications have incorporated "action-at-a-distance" in their neighbourhood definitions. For instance, while modelling the evolution of urban land-use types, White and Engelen (1993) defined quite a large neighbourhood size comprising 113 cells within a circle of a six-cell radius (Figure 2.7a). The distance-decay effect was applied to these neighbouring cells: the further the distance from the central cell in question, the less the influence these neighbouring cells had on the central cell. The city grew, and its structure evolved as cells were converted from one state to another according to the transition rule, which was a function both of current land uses in the 113 neighbouring cells and of the inherent suitability of the cell for each possible land use.

To simulate the land development in a fast growing region in China using a linguistic cellular automata model, Wu (1996) employed a rectangular neighbourhood of a 5×5 cell, which included 120 cells surrounding the cell in question (Figure 2.7b). Although it is commonly agreed that influence may be produced by cells at a distance, the problems of how far away these cells are in order to influence the central cell and how important this influence is to the transition of states remain uncertain, and it requires further research.

2.3.3.2 Neighbourhood Size

Similar to the construction of cell scales, the selection of the neighbourhood size is also pragmatic. According to the theory of cellular automata, the global behaviour of a self-organising system is governed by locally defined transition rules. For an urban system, a fundamental question is to what extent urban development is a locally specified process (Wu 1996). Some factors, such as slope aspect and height of land, affect urban development in a small area base; others such as urban planning and the transportation networks are global controls over the whole area. Moreover, developments in information technology and telecommunications have had fundamental consequences for the patterns and processes of urban change throughout the world (Herbert and Thomas 1997). These factors affect urban development in a universal way.

In practice, both small and large neighbourhood sizes have been applied to models of urban development. For instance, Clarke and Gaydos (1998), Wu (1998a,b,c, 1996) and Clarke, Hoppen, and Gaydos (1997) used the Moore Neighbourhood, which is small, as it only consists of nine cells including the cell in question. As for White and Engelen (1997, 1994, 1993), the impact of the 113 neighbouring cells on the central cell in question varies based on the "distance-decay" rule, that is, cells that are closer are weighted more heavily, whereas cells that are further away may carry light or no weight toward the transition potentials. In addition, the weights may also be positive, representing an attractive effect, or negative if two states are incompatible (White and Engelen 2000; Engelen et al. 1999; White 1998).

However, no particular validation on the size of the neighbourhood in cellular automata-based urban models has been explored. Most applications of cellular automata models in urban research employ a larger neighbourhood size than applications in the natural sciences (Batty and Xie 1994c). This is probably because of the difficulty in justifying transition rules in behavioural terms (Wu 1996) and the existence of distance-decay effects of the neighbouring cells to the central cell in question (Wu 1996; White and Engelen 1994, 1993).

2.3.3.3 Neighbourhood Type

Regardless of the size of the neighbourhood, the type of neighbourhood also has significant impacts on the behaviour of a cellular automaton. Li and Yeh (2000) show that the use of a rectangular neighbourhood, such as the Moore Neighbourhood, might produce significant distortions between cells at different directions from a circular object. In fact, owing to the distance-decay effect of the neighbouring cells on the central cell in question, the application of a rectangular neighbourhood in a cellular automata model can produce distortion on an object of any shape (Figure 2.8).

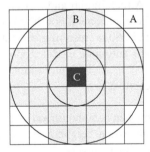

FIGURE 2.8 Distortion produced by a rectangular neighbourhood. (C (dark colour) is the processing cell. For a rectangular neighbourhood of 7×7 cells, the effect of cell A (one of the white-coloured cells) on the processing cell (C) differs from that of cell B (one of the grey-coloured cells), although both A and B are in the same row of the neighbourhood. This is because the distances from the centres of A or B to the processing cell C are different.)

This distortion is especially significant when a large neighbourhood size applies, which can be eliminated by applying a circular neighbourhood.

2.3.3.4 Irregular Neighbourhood

Similar to the use of irregular cell scales, the literature also shows the application of an irregular neighbourhood scale in cellular automata-based urban models. For instance, in the Voronoi-based cellular automata model developed by Shi and Pang (2000), the neighbourhood objects are defined as those that share common Voronoi boundaries to the spatial object in question within a certain distance. The influence of the neighbouring objects can be defined as a function of the distance between two spatial objects, and a "distance-decay" effect can be reflected in the transition rules (Shi and Pang 2000: 462).

In the cellular automata model developed by Stevens and Dragicevic (2007), three irregular cellular neighbourhoods were proposed and implemented into their iCity model. The adjacency neighbourhood includes all polygons that share a common edge or point with the polygon in question; the distance neighbourhood includes all polygons that fall completely or partially within a certain distance of the polygon in question; and the clipped distance neighbourhood includes all polygons that fall within a certain distance of the polygon in question plus the portions of polygons partially within the same distance of the polygon. Compared to the cell based regular neighbourhood type, the irregular neighbourhoods suit well to land parcel proximity functions (Stevens and Dragicevic 2007).

2.3.3.5 Sensitivity Analysis

Owing to the significant impact of neighbourhood scale and type on the behaviour and outcomes of cellular automata models, some research literature reports on their sensitivity in relation to the spatial scale and neighbourhood configurations. For instance, Kocabas and Dragicevic (2006) used the cross-tabulation map, Kappa index with coincidence matrices and spatial metrics, to systematically evaluate the impact of the neighbourhood size and type on the behaviour and outcomes of the cellular automata model. A rectangular and a circular neighbourhood type were selected with 9 simple and 45 complex neighbourhood configurations, each at spatial resolutions of 50, 100, 150, and 250 m, resulting in a total of 432 simulations. Their results suggest that the cellular automata model responds differently depending on the spatial resolution as well as the neighbourhood size and type. Specifically, the cellular automata model is sensitive to

1. Changes in spatial scale when neighbourhood size and type are kept constant
2. Changes in spatial scale when neighbourhood size is kept constant but neighbourhood type is being changed
3. Changes of neighbourhood size when spatial scale and the neighbourhood type are kept constant
4. Changes of neighbourhood type when spatial scale is kept constant but the neighbourhood size is being changed (Kocabas and Dragicevic 2006: 950)

Other research by Menard and Marceau (2005) also reported the combined effects of the spatial scales of cells and neighbourhood. Their research demonstrated that, although the simulation results of a geographical cellular automata model are sensitive to both the cell size and neighbourhood scale, the choice of a neighbourhood scale is less influential on simulation results; it becomes more sensitive only when a large-scale neighbourhood (e.g., a circular neighbourhood of approximately a five-cell radius) is used, or when a medium-sized neighbourhood (e.g., a circular neighbourhood of an approximately two-cell radius) is applied in combination with a large cell size of 1000 m (Menard and Marceau 2005).

However, the mechanism underpinning the selection of the neighbourhood scales and the configuration of a neighbourhood scale with which the cellular automata models can generate scenarios that best mimic the actual process of urban development still need to be addressed.

2.3.4 Variation in Transition Rules

The core component of a cellular automata model is the transition rules that represent the logic of the process being modelled; therefore, they determine the spatial dynamics of the system (White and Engelen 2000). The fundamental role of the transition rules is to serve as the algorithms that drive the change of cells from one state to another over time.

Various approaches have been employed in defining the transition rules of an urban cellular automata model based on the understanding of an urban system and its evolution from different perspectives, resulting in different types of urban cellular automata models. These approaches range from the very simple to the very complex ones. For instance, the diffusion-limited aggregation model developed by Batty and colleagues for modelling the dynamic urban growth is a cellular automaton in nature, and the rule applied in this model is simple: a vacant cell would convert to an occupied one if it is within the neighbourhood of an occupied cell (Batty, Longley, and Fortheringham 1989). However, other urban models based on cellular automata use combined rules to simulate the complex behaviour of the system. This section reviews urban cellular automata models in the literature with distinct features in defining the transition rules.

2.3.4.1 Constrained Cellular Automata

The contemporary cellular automata model of urban growth started with research by White and Engelen, who first developed and applied a constrained cellular automata model to simulate land-use change dynamics (White and Engelen 1993). Unlike the standard cellular automata where the development of cells in each state is determined endogenously by the transition rules, their model took into account the constraints of other factors, including the qualities of the land, the effects of neighbouring land-use activities, and the aggregate level of demand for each category of land. Therefore, their model consists of two parts: a macroscale model and a microscale model. The macroscale model was developed exogenously to the cellular automata model based on factors such as population and socio-economic status.

This model generates constraints with different land-use consumptions that were applied to control the amount of cells that must be in a certain state, so that the model can achieve a realistic representation of the system being modelled.

At the micro scale, a set of transition potentials representing the inherent suitability of a cell from its current state into other land uses was estimated. The transition potentials were calculated as a function of a number of factors including the accessibility of the cell to the road network, the intrinsic suitability of the land, the zoning status, and the impact of the neighbourhood on the cell for a particular land use. All cells are ranked by their highest transition potentials, and the state of each cell will be changed to the state for which it has the highest potential. However, the number of cells in each state is determined or constrained by the macroscale model; hence the name constrained cellular automata model. The transition of cells begins with the highest ranked cell and proceeds downward. When a sufficient number of cells of a particular land use have been achieved, the potentials for that land use are subsequently ignored in determining cell transitions. Therefore, some cells may not be in the state for which they have the highest potential; these cells will remain in their current states. Such a transition rule is applied to all cells at all iterations (Engelen, White, and Uljee 2001; White and Engelen 2000).

The constrained cellular automata model provides a framework for research that integrates cellular automata with other socio-economic and environmental models, and that operates at both micro and macro geographical scales. It also provides a logical link between the traditional top-down approach applied in urban modelling and the current bottom-up approach based on cellular automata theory. The constrained cellular automata model has evolved and been applied to simulate land-use change in a number of cities since its first conceptualisation in 1993 (White and Engelen 2000; White and Engelen 1997; Engelen, White, and Uljee 1997; White, Engelen, and Uljee 1997; Engelen et al. 1995; White and Engelen 1993).

2.3.4.2   The SLEUTH Model

Clarke and colleagues (Clarke, Hoppen, and Gaydos 1997) developed a model to simulate the process of urban growth based on cellular automata. The name of the model came from the six input data layers, namely Slope, Land cover, Exclusion, Urbanisation, Transportation, and Hillshade. Hence, the model is named SLEUTH. The model captures urban patterns through the application of four types of urban land-use change: spontaneous growth, new spreading centre growth, edge growth, and road-influenced growth. A *spontaneous growth* occurs when a randomly chosen cell falls close enough to an urbanised cell, simulating the influence of urban areas on their surrounding land; a *new spreading centre growth* spreads outward from existing urban centres, representing the tendency of cities to expand; an *edge growth* urbanises cells that are flat enough to be desirable locations for development even if they do not lie near an already established urban area; and a *road-influenced growth* encourages urbanised cells to develop along the road network.

The four types of urban growth are applied sequentially during each growth cycle and are controlled through the interactions of five growth coefficients: diffusion, breed, spread, road gravity, and slope (Clarke, Hoppen, and Gaydos 1997; Clarke and Gaydos 1998). The first four coefficients describe the growth pressure in the urban system. For instance, the diffusion coefficient determines the overall outward dispersive nature of the distribution; the breed coefficient specifies how likely a newly generated detached settlement is to begin its own growth cycle; the spread coefficient controls how much diffusion expansion occurs from existing settlements; and the road-gravity factor denotes the attraction of new settlements toward and along roads. Resistance to growth is incorporated through the slope coefficient, which captures the effect of steep slopes on restricting development. In addition, resistance is also applied through an excluded data layer that identifies areas that are wholly (e.g., water or parks) or partially (e.g., restrictive zoning) excluded from development. All five coefficients are calibrated to control the growth rate so that growth will not become unusually high or low. The overall rate of urban growth is the sum of the four types of growth.

The SLEUTH model is implemented in two general phases: a calibration phase, in which the model is trained to replicate historic development trends and patterns, and a prediction phase, in which historic trends are projected into the future. The model can be used to simulate the non-urban to urban conversion; it can also model the process of multiple land-use change. The model was first developed and applied in the San Francisco Bay area in the United States (Clarke, Hoppen, and Gaydos 1997) and subsequently to predict urban growth in San Francisco and the Washington/Baltimore region (Clarke and Gaydos 1998). Since then, the model has been calibrated and widely used to model urban growth throughout the various regions of the United States and the world (Dietzel and Clarke 2006, 2004; Syphard, Clarke, and Franklin 2005; Jantz, Goetz, and Shelley 2004; Leão, Bishop, and Evans 2004; Yang and Lo 2003; Silva and Clarke 2002).

2.3.4.3 Fuzzy Constrained Cellular Automata Models

Apart from defining transition rules with deterministic specifications, research also shows that transition rules of a cellular automaton may not necessarily be restricted to deterministic forms. More flexibility in defining the rules, such as the application of probability concepts and fuzzy logic, has been tested, implying that the complex system behaviour can be simulated through the appropriate definition of the transition rules based on fuzzy set theory (Liu and Phinn 2003; Wu 1996; White and Engelen 1993). As urban development is the result of both physical constraints and human decision-making behaviour, which are both characterised by uncertainty and fuzziness, the applications of the fuzzy set theory and fuzzy logic control seem attractive in defining the rules controlling urban developments.

A number of studies on the application of fuzzy set and fuzzy logic in cellular automata-based urban modelling can be identified in the literature (Mandelas, Hatzichristos, and Prastacos 2007; Dragicevic 2004; Liu and Phinn 2003; Wu 1998b) following the pioneering attempt by Wu (1996). A linguistic cellular automata simulation approach was developed by Wu (1996) to simulate the process of land development in a fast growing urban region in China. In this simulation approach, the transition rules are defined to integrate with a decision-making process, that is, they are defined through vague and subjective natural language statements to describe certain preconditions of a decision. For instance, the development of a piece of agricultural land into urban land use is most likely preconditioned by having the quality of *good* accessibility. The rule can be easily modified with linguistic hedges such as *very good* or *not so good*. The natural language statements allow the integration of fuzzy logic control into mimicking the decision-making process, thus making the structure of the model more transparent and comprehensible (Wu 1996: 368). Application of the model in China's Guangzhou City shows that the model can simulate urban development in a gaming style. It is a useful tool to provide enlightening scenarios to decision makers. However, as the model was configured at a local scale, that is, at the scale of cells and its neighbourhood, it did not take into account the impact of macroscale factors such as land-use zoning and land availability on urban development. Moreover, as the membership function and the fuzzy linguistic modifiers are defined in a subjective way, the interpretation of the model's results is largely restricted (Wu 1996).

As the focus of this book, more discussions on the fuzzy constrained cellular automata model will be presented in the following chapters of the book.

2.3.4.4 Transition Rules Derived from Other Models

Another popular approach used in cellular automata-based urban modelling is to incorporate other modelling methodology into the model, especially for defining the transition rules for it. Mathematical models used include the analytical hierarchy process (AHP), an approach of the multicriteria evaluation (Wu and Webster 1998; Wu 1998c), multiple regression analysis (Sui and Zeng 2001), principal components analysis (PCA) (Li and Yeh 2002), and so forth. For instance, in Wu (1998c) and Wu and Webster (1998), a model called SimLand was developed, which integrates GIS, cellular automata, and AHP as three interrelated components. The AHP method was used to derive behaviour-oriented rules of transition; it uses pairwise comparisons to acquire the preference of decision makers. The AHP approach was developed by Saaty (1980), and it has been implemented in a GIS environment as well (Banai 1993). Results from the AHP were used as an indicator of land conversion probability to feed the cellular automata model in configuring the transition rules. GIS was used as a modelling framework and also to present and visualise simulation results. The model was also applied to simulate the land development of an urban region in China's Guangzhou City (Wu 1998).

In Sui and Zeng (2001), a multiple regression method was developed to calculate weights to be used in the cellular automata model. Parameters used in the multiple regression method were computed in three consecutive steps. The first step was to overlay all temporal data from 1992 to 1996 in order to identify and extract cells that have changed states over that time period. The second step was to use a moving window of 10×10 cells to select cells that were converted into urban land. The proportion of converted cells in each moving window was taken as the transition probabilities for cells in that window. The last step was to measure a mean value of factors such as elevation for cells within the window. This model was applied to simulate the dynamics of landscape structure in a fast growing urban area in southeast China (Sui and Zeng 2001).

In Li and Yeh (2000), similar to the constrained cellular automata models developed by White and Engelen (1993, 1997), multiple scale constraints were identified at local, regional, and global scales. The local constraints were represented by cell-based values or scores that influence the cellular automata model at the local or cell scale. The regional constraints emphasize the differences of a geographical phenomenon among larger areas. These constraints such as administrative boundaries or local government planning policies may only have aggregated or partial-spatial information to be used in the model. The global constraints are normally characterised by temporal or non-spatial information, but they can be used to control the overall amount of land consumption in the cellular automata modelling process (Li and Yeh 2000). The various constraints were incorporated through multiple criteria evaluation techniques to generate development suitability for the cellular automata-based urban model. Subsequently, Li and Yeh (2002) also used principal components analysis (PCA) to deal with the correlation of spatial variables and remove redundant data.

In the effort of using various mathematical approaches to configure the transition rules of a cellular automaton, it is obvious that the primary purpose of employing mathematical approaches in the modelling practice is to evaluate the suitability or probability of land for development. Therefore, this type of model is also regarded as a "suitability based CA model" (Li and Yeh 2000).

2.3.4.5 Artificial Neural Network (ANN)-Based Cellular Automata Models

Either the constrained cellular automata model, or the SLEUTH model, or the suitability-based cellular automata models require the configuration of parameter values for the transition rules, which are critical to the model. Although it is not always easy to develop detailed transition rules with correctly configured and calibrated parameter values in a cellular automata model, an approach based on ANN was developed and tested by Li and Yeh (2002, 2001) to generate and calibrate the model's parameter values automatically. ANN is a mathematical model based on biological neural networks. It consists of an interconnected group of artificial neurons, and it processes information using a connectionist approach to computation. The neural network contains three types of data layers: input data layers, output data layers, and some hidden layers staying in between the input and output layers (Figure 2.9). Similar to the human brain, the ANN model can be trained by sample data to learn, think, and react to stimulus. During this training process, the initial model parameter values are modified repeatedly until the model can generate acceptable outcomes that match the targeted output values.

Using ANN, Li and Yeh (2002, 2001) formalised a model to incorporate in the transition functions of the cellular automata. A three-layer neural network with multiple output neurons was designed to calculate conversion probabilities for multiple land uses. The ANN was trained using data extracted from the local site in GIS. Through a set of training processes, the neural network was able to generate

Input Layers   Hidden Layers   Output Layers

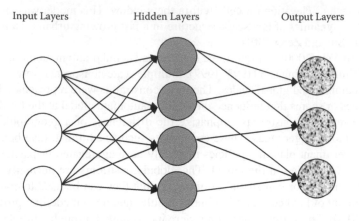

FIGURE 2.9 An artificial neural network.

a number of parameter values automatically, which were imported into the cellular automata model to simulate the multiple land-use change progress. Their study shows that the ANN-based cellular automata model was able to generate results with better accuracy than a non-ANN based model in the simulation of nonlinear complex urban systems (Li and Yeh 2002, 2001). However, a fundamental issue in using the neural network approach is that it is essentially a black-box type of model; what happens inside the black box is unknown to the modellers and the users. Therefore, the model does not provide explicit knowledge in understanding the process of land-use change (Li and Yeh 2002).

2.3.4.6 Stochastic Cellular Automata Model

Research literature also shows the use of a stochastic approach in configuring the parameter values of a cellular automata model. One such study was undertaken by Wu (2002) who derived the initial probability of simulation from observed sequential land-use data. The initial probability was updated dynamically through local rules based on the strength of neighbourhood development. By applying the model to simulate the process of rural to urban land use conversion in China's Guangzhou city, the model generated realistic simulation results.

Similarly, Almeida et al. (2003) designed a framework using probabilistic methods that are based on Bayes' theory and the related "weights of evidence" approach for simulating urban change (Almeida et al. 2003: 481). The model was applied to a medium-sized town in Brazil to combine various socio-economic and infrastructural factors to predict the probability of changes between land-use types. The research shows that the "weights of evidence" approach provides a particularly simple and useful way of illustrating how less conventional map data in raster and in binary form can be used in a multivariate framework, thus linking cellular automata and raster GIS to more conventional and well-established methods of statistical estimation. An advantageous feature of the model is that factors that are directly considered by local municipalities and developers who have the greatest control over urban land development are used as drivers of the growth process (Almeida et al. 2003: 507).

2.3.5 MODELLING TIME

Urban development is a process that occurs in space and over time. In order to understand the dynamic process of such development, both the spatial and temporal dimensions of this process need to be taken into account. Many efforts have been made at the conceptual level to improve the temporal dynamic representation in a GIS (Claramunt and Theriault 1995; Peuquet and Duan 1995; Peuquet 1994; Langran 1993; Hazelton, Leahy, and Williamson 1992). However, the paradigm for temporal data representation inside a GIS framework still remains unresolved (Dragicevic 2000).

According to Snodgrass (1992), there are several definitions of time in the context of a GIS database. These include

1. The *real world time* when change really occurs
2. The *updating time* when geographical data are recorded
3. The *cartographic time* when data products are released
4. The *database (transaction) time* when data are registered into the database (Snodgrass 1992)

More recently, Dragicevic (2000) developed an approach based on fuzzy set theory to generate new data at times intermediate to existing data layers. The model could closely simulate the evolution occurring in the real world given that an optimal temporal resolution is chosen. Consequently, the database representation of time can approach the real-world continuum (Dragicevic 2000).

For cellular automata models, one of the fundamental components is time, that is, the change of state of a cell at one time is controlled by the state of the cell itself and the state of cells within a certain neighbourhood in the previous time step. Hence, the model needs to be configured not only spatially but also temporally. However, current practices in cellular automata-based urban modelling are mainly focused on the spatial dimension of urban land-use change; there is little concern for the temporal dimension of the model. Most cellular automata models are configured by starting the model from a certain time where spatial data sets are available, and then letting the model run for a number of iterations until the simulated results "fit" with another set of data at the ending time. Therefore, these models were not configured temporally; they were not designed to simulate the process of urban development over time.

2.4 CONCLUSION

This chapter introduced the basic cellular automata model, its origin in computing, and its applications in urban modelling practices. A growing number of models are developed with a view to representing real urban and regional systems and their use for practical tasks, resulting in significant progress in the field of simulating the spatial and temporal dynamics of urban development. However, there are still several unanswered questions remaining in this field.

- Most of the cellular automata models of urban growth configure the cells as binary states of either urban or non-urban, or with specific land-use types; there is a sharp boundary between the non-urban and urban states, or between the different types of land use. Therefore, those models simulate only the conversion of a non-urban to an urban process (Li and Yeh 2000; Clarke and Gaydos 1998; Wu 1998a,b,c, 1996; Clarke, Hoppen, and Gaydos 1997).
- Although some research has been undertaken to apply fuzzy set theory in defining the transition rules of a cellular automata model (Wu 1998b, 1996; Dragicevic 2004), more systematic research needs to be done to study the fuzzy rule transition mechanism of an urban fuzzy set; the fuzzy transition rules should also be associated with the non-determinative or fuzzy nature of the cell state.
- The configuration of the neighbourhood is pragmatic. No specific consideration has been given to the setting of the neighbourhood size in terms of either theory or practice.
- Most of the cellular automata models are not configured temporally. These models are implemented by setting up a starting date and then letting the model run for a number of iterations until the simulated results "fit" with the data set for calibration. Consequently, they only model the spatial process of urban development.

This book will address some of these issues, if not all of them. The following chapters present a fuzzy constrained cellular automata model of urban development and apply the model to simulate the spatial and temporal processes of the urban development of a metropolitan region in Sydney, Australia.

3 Developing a Fuzzy Constrained Cellular Automata Model of Urban Development

Urban development is the result of a variety of factors, these being physical, socio-economic, and institutional. Previous studies on urban modelling have addressed various aspects of urban development. However, most of these studies regard urban development as a binary process of non-urban to urban conversion conducted under the paradigm of the crisp set theory (Wu and Webster 2000; Clarke and Gaydos 1998; Wu 1998a,b,c, 1996; White and Engelen 1997, 1994, 1993). In fact, the process of urban development resembles a fuzzy process both spatially and temporally. Spatially, there is no sharp boundary between an urban built-up area, urban–rural fringe, and non-urban rural land. Temporally, urban development is a continuous process that can be illustrated using a logistic curve (Herbert and Thomas 1997; Jakobson and Prakash 1971).

This chapter develops a fuzzy constrained cellular automata model of urban development. Urban areas are defined based on a fuzzy set approach. A fuzzy membership function is used to delimitate the continuous process of the rural-to-urban transition. In addition, the transition of cells from one state to another is controlled by fuzzy logic-constrained rules representing the non-deterministic nature of forces leading to urban development. Section 3.1 introduces fuzzy set theory and the role of fuzzy set in defining the state of cells of an urban cellular automaton. Section 3.2 discusses fuzzy logic and fuzzy logic control. It also demonstrates how a fuzzy logic controller can be introduced to model the process of urban development. This is followed by Section 3.3, which illustrates the construction of a fuzzy logic-controlled cellular automata to model the process of urban development. A number of primary and secondary rules as well as a rule-inferencing engine are proposed based on fuzzy logic; these rules and the fuzzy inferencing engine are calibrated when the model is applied to a real urban extent. Section 3.3 also presents a discussion on the defuzzification approach used to transfer the model's fuzzy output results into crisp values. Finally, conclusions on model construction based on fuzzy set and fuzzy logic are presented in Section 3.4.

3.1 URBAN DEVELOPMENT AND FUZZY SETS

Urban development is a process of physical concentration of people and buildings (Herbert and Thomas 1997). This is a continuous process in space and time that resembles a fuzzy process in both its definition of urban areas and the rules or factors controlling the development. Although fuzzy set and fuzzy logic approaches have been developed since the 1960s, in practice, the understanding of an urban development as a probabilistic and fuzzy process remains in the realm of assertions. The fuzzy characteristics of urban development imply the applicability of the fuzzy set approach in modelling the process of this development.

3.1.1 FUZZY REPRESENTATION OF GEOGRAPHICAL BOUNDARIES

Traditionally, it is common to use thematic maps to represent geographical phenomena (Woodcock and Gopal 2000; Wang and Hall 1996). Here, a thematic map usually refers to a collection of spatial entities that are defined "both by their location in space and by their non-spatial descriptions about a theme" (Wang and Hall 1996: 573). In thematic maps, the use of categories has followed the classical set theory, where each location is assumed to belong to a single category, and the boundaries between different categories are represented as sharp lines (Woodcock and Gopal 2000; Burrough 1986). This representation might be accurate when dealing with cadastral, census, or administrative boundaries that are "sharply defined" (Wang and Hall 1996: 574). Unfortunately, it is not accurate in representing the boundaries of land features with continuously changing properties, such as soil quality, land cover, or population densities because such boundaries are rarely sharp or crisp. In this case, the representation of geographical boundaries based on the crisp set theory may lead to misunderstanding of the information being represented (Wang and Hall 1996).

As in many dynamic processes of geographical phenomena, urban development is a continuous diffusion process in space and over time. Spatially, an urban area is an area with a high concentration of residential land use and the dominance of non-agricultural land. This is quite a fuzzy definition, and there is a variety of differences in determining either the degree of concentration of resident population or the extent of the dominance of non-agricultural land for an area to be regarded as urban. Moreover, all cities are surrounded by rural or natural land; there is no sharp boundary between an urban built-up area and its non-urban hinterland. Between the well-recognised urban land use and the area devoted to agriculture, there exists "a zone of transition in land use, social and demographic characteristics, lying between (a) the continuously built-up area and suburban areas of the central city, and (b) the rural hinterland, characterised by the almost complete absence of non-farm dwellings, occupations and land use" (Pryor 1968: 206). This "zone of transition" is a place where both urban and non-urban features occur, which has been broadly termed as *fringe* or *rural–urban fringe* (Bryant, Russwurm, and McLellan 1982; Pryor 1968). This fringe area has become the most vigorous part of development in the rural–urban continuum and has attracted much attention in urban research. However, different concepts or terminology have been applied in the literature, such as "suburb," "fringe," "urban fringe," "rural fringe," "inner fringe," "urban shadow zone," "exurban zone," or even "rurban fringe" (Bryant, Russwurm, and McLellan 1982; Carter 1995; Kurtz and Eicher 1958; Wehrwein 1942). This is largely due to the complexity of its spatial form and socio-economic features, the constantly changing nature of development, as well as the diverse extent of development in space. Obviously, the terminology used is imprecise, thus making the study of the spatial structure of urban systems very complex and less comparable (Kurtz and Eicher 1958).

Temporally, if an area has been developed from one state (non-urban) to another (urban) within a certain period of time, development has actually taken place continuously within this period. Therefore, it is difficult and also inaccurate to define an exact time spot when the actual change of state occurred.

The development of the fuzzy set theory and its application for representing geographical phenomena has provided a solution for dealing with such problems. Unlike the crisp set theory where a location in a landscape either belongs to a map category exclusively or it does not belong to it at all, fuzzy set theory allows partial belongings represented by a grade of membership in this fuzzy set. In the following section, the terminology of the fuzzy set theory and its application in delimiting urban areas is presented. With this terminology, a method of defining urban states based on fuzzy set for modelling urban development with cellular automata is pursued in Section 3.1.3.

3.1.2 FUZZY SET THEORY

3.1.2.1 Definition of Fuzzy Set

In set theory, a classical (crisp) set is normally defined as a collection of elements or objects that can be finite, countable, or overcountable. Each single element can either belong to or not belong to the set, based on Boolean logic (Whitesitt 1961). There is no overlap between class memberships. According to the crisp set theory, an urban area can be defined as an entity bounded by a sharp boundary that separates it from its non-urban counterparts. The inclusion or exclusion of an area within one class is usually decided based on some chosen criteria. However, such criteria may be difficult to define in reality, and they often require the application of the probability theory (Burrough and Frank 1995).

The fuzzy set theory was developed in the 1960s by Zadeh (1971, 1965, 1962). This theory was formulated to extend the crisp set theory in order to deal with continuous classifications. A set is fuzzy if an element can belong partly to it, rather than having to belong to the set completely or not at all. Therefore, the fuzzy set theory begins with the assignment of membership grades to elements that are not restricted to 0 (non-membership) or 1 (full membership), but which may lie somewhere in the interval from 0 to 1 (partial membership). Mathematically, a fuzzy set can be expressed as follows.

Let X be a collection of objects, whose generic element is denoted as x. Thus $X = \{x\}$. A fuzzy set A in X is a set of ordered pairs,

$$A = \{(x, \mu_A(x)) \mid x \in X\} \tag{3.1}$$

where $\mu_A(x)$ represents the grade of membership of x in A, which associates with each x a real number in [0,1].

Consider a city in a regional context. Within this region, some areas have been fully developed as urban built-up areas, such as the central business district, the highly populated residential areas, or the concentrated industrial zones. Some areas remain in a non-urban state, such as the surrounding open or agricultural land or the regional recreational land. These areas are identified as non-urban or rural areas. Except for these two extreme categories, there are also areas that may not have been fully developed, such as areas with low-to-medium population density or with a mixture of agricultural and industrial land uses. These areas can be categorised as partially urban areas.

The temporal dimension of this development is similar to that of its spatial dimension. If an area has been developed from non-urban to an urban built-up area within a certain period, the development has been occurring as a continuous process over this period of time. Therefore, within this time span, the area could have been partially developed to some extent. To illustrate the extent of development of this city in space and over time, a fuzzy concept of "urban" or "non-urban" can be defined based on the fuzzy set theory.

Let X be a collection of cells representing an area in a regional context. x_{ij} is a generic form of a cell in X. An urban fuzzy set S_{urban} can be defined as a set of ordered pairs,

$$S_{urban} = \{(x_{ij}, \mu_{S_{urban}}(x_{ij}))|\ x_{ij} \in X\} \tag{3.2}$$

where the $\mu_{S_{urban}}(x_{ij})$ is a membership function, indicating the extent of the cell x_{ij} belonging to the urban fuzzy set S_{urban}. This membership function represents the state of the cell x_{ij} undergoing an urban development process.

Similarly, a non-urban fuzzy set, $S_{non-urban}$, can also be defined as

$$S_{non-urban} = \{(x_{ij}, \mu_{S_{non-urban}}(x_{ij}))|\ x_{ij} \in X\} \tag{3.3}$$

where the $\mu_{S_{non-urban}}(x_{ij})$ is another membership function, indicating the extent of the cell x_{ij} belonging to the non-urban fuzzy set $S_{non-urban}$. The value of this membership function represents the degree or extent to which the cell remains in the non-urban fuzzy set.

3.1.2.2 Membership Function

The membership function determines how and to what degree a cell belongs to the set. It depends on the extent to which a cell is being developed in the urban growth process. The closer the grade of membership is to 1, the higher the degree of membership of the cell in that fuzzy set. Figure 3.1 illustrates two linear membership functions, one for an urban fuzzy set and the other for a non-urban fuzzy set. For instance, if a cell has a membership grade of 0.7 in the urban fuzzy set, this cell would have been developed to a higher degree than a cell with a membership grade of 0.3 in the same urban fuzzy set. Conversely, a cell with a membership grade of 0.8 in the non-urban fuzzy set would have been less developed compared to a cell with a membership grade of 0.4 in the same non-urban fuzzy set. With this terminology, the boundary between non-urban and urban areas can be understood not as a sharp line but as a region with continuous change on the scale of membership.

Obviously, the membership function is a crucial component of a fuzzy set. Different membership functions represent different fuzzy sets, even though they may have a

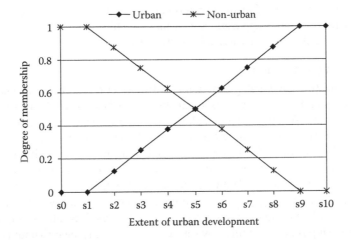

FIGURE 3.1 Two linear membership functions, one for an urban and one for a non-urban fuzzy set. (The various stages of urban development are represented by s0 to s10, ranging from not developed to fully developed.)

similar context. Figure 3.2 illustrates three membership functions, one using a linear function, one using an exponential function, and one using a logarithmic function. Each function represents a different fuzzy set although they all have similar contexts, that is, "a class of cells that has been developed in the urbanisation processes."

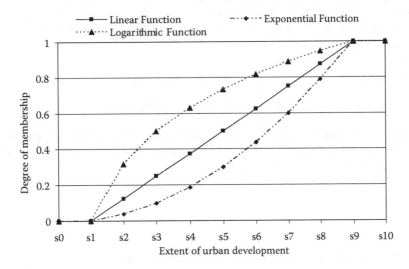

FIGURE 3.2 Variations of fuzzy membership functions. (The various stages of urban development are represented by s0 to s10. The three membership functions represented in this figure define three different fuzzy sets although they have similar content, that is, before Stage s1, the area is undeveloped, and after Stage s9, the area is fully developed. Development has taken place in areas with an urban extent between s1 and s9. However, their membership grades vary and are determined by their membership functions.)

To determine what type of membership function is suitable for a set depends on the context of the particular application. The membership function defining the urban extent in this book will be discussed in Section 3.1.3.2.

3.1.2.3 Fuzzy Operation

Not only does the membership function of a fuzzy set represent the property of the set, logical operations based on Boolean logic can also be applied to fuzzy sets; their membership functions can be calculated using Boolean algebra. For instance, the binary operators of logical OR and logical AND can be applied to perform the Union and Intersection of two fuzzy sets. A Union of two sets using the Boolean operator "OR" results in a new set that contains everything that belongs to either of the two sets, but nothing else. With the Union operation, an urban fuzzy set S_{urban} and a non-urban fuzzy set $S_{non-urban}$ can be combined to generate a new set, which can be termed as an urban or non-urban set $S_{urban\_or\_non-urban}$, as illustrated in Equation 3.4. The membership function of this new set, $\mu_{S_{urban\_or\_non-urban}}$, is the maximum of the two membership functions in the original urban or non-urban fuzzy sets, which can be represented in Equation 3.5.

Union
$$S_{urban\_or\_non-urban} = S_{urban} \cup S_{non-urban} \tag{3.4}$$

$$\mu_{S_{urban\_or\_non-urban}}(x_{ij}) = \max(\mu_{S_{urban}}(x_{ij}), \mu_{S_{non-urban}}(x_{ij})) \tag{3.5}$$

Similarly, using the Boolean operator "AND," an Intersection of two fuzzy sets can be generated to create a new fuzzy set $S_{intersection}$, which can be represented in Equation 3.6. The membership of this new set, $\mu_{S_{intersection}}$, is therefore the minimum of the memberships for the S_{urban} and $S_{non-urban}$ fuzzy sets, which can be represented in Equation 3.7.

Intersection
$$S_{intersection} = S_{urban} \cap S_{non-urban} \tag{3.6}$$

$$\mu_{S_{intersectionn}}(x_{ij}) = \min(\mu_{S_{urban}}(x_{ij}), \mu_{S_{non-urban}}(x_{ij})) \tag{3.7}$$

In addition, the unary operator of logical NOT can also be applied to generate the complement of a fuzzy set, $\overline{S}_{urban}$, as can be represented in Equation 3.8; its membership function is defined in Equation 3.9.

Complement
$$S_{complement} = \overline{S}_{urban} \tag{3.8}$$

$$\mu_{S_{complement}}(x_{ij}) = 1 - \mu_{S_{urban}}(x_{ij}) \tag{3.9}$$

With this example, the intersection of urban and non-urban fuzzy sets, $S_{intersection}$ can be considered as a set that belongs to both urban and non-urban fuzzy sets. This can be best described as a partly urban set that has both urban and non-urban characteristics. The union of the two fuzzy sets, $S_{urban\_or\_non-urban}$, is a set of either urban or non-urban, where a cell receives a higher grade of membership if it has more urban or non-urban features, and it receives a lower grade of membership if it is featured more likely as a partly urbanised area (Figure 3.3).

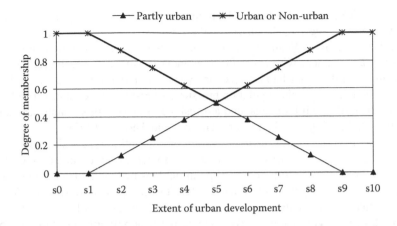

FIGURE 3.3 Membership functions of the intersection and union of two fuzzy sets. (The different extents of urban development are represented by s0 through s10, with s0 being a lower extent and s10 a higher one.)

Similarly, the complement of the urban fuzzy set $S_{complement}$ is also a fuzzy set, which can be identified as a non-urban fuzzy set.

3.1.3 URBAN DEVELOPMENT AS A FUZZY PROCESS

3.1.3.1 Defining Urban Areas

A number of approaches have been applied to measure the extent of urban development. These include the use of detailed rules of the size and density of the population for the definition of urban areas, the use of population densities, and the use of remote sensing technology to identify and measure land-use types. As cities are physical agglomerations of population and housing, there should be some kind of threshold measurements along the population-size continuum of settlements at which a village becomes a town, although this threshold may vary significantly in space and time (Herbert and Thomas 1997).

Various criteria on the size and density of population have been found in defining the urban areas by different countries around the world. For instance, in several Scandinavian countries, any settlement that has more than 200 inhabitants is classified as urban in the national census (Herbert and Thomas 1997); in the United States, the "urban areas" comprise places with 2,500 or more inhabitants, and some special types of areas having a density of 1,500 inhabitants per square mile (about 580 inhabitants per square kilometre) (U.S. Census Bureau 1960). In Canada, an urban area should have a minimum population concentration of 1,000 and a population density of at least 400 persons per square kilometre based on the previous census population counts (Statistics Canada 1996). In Australia, areas with a population density of 500 persons per square mile are classified as urban; however, there are some considerations for modifying this criterion (Linge 1965). Many countries impose much higher thresholds of population size, such as Greece, with 10,000 inhabitants, and Japan with 30,000 (Herbert and Thomas 1997). The diversity of this threshold largely relates to the social context of the country. For instance, considering the physical geography of Scandinavia and the ways in which its settlements evolved over time, a settlement with over 200 permanent inhabitants is well regarded as an urban area. On the other hand, in a country such as Japan with a relatively limited land area and considerable population pressure, almost all settlements exceed such a low threshold of 200 inhabitants, and a threshold of 30,000 inhabitants seems more realistic in delimiting its urban extent (Herbert and Thomas 1997).

In addition to the use of the combination of population size and density, a pure criterion of population density was also applied in delimiting urban areas. Gryztzell (1963) presented a method based on population densities alone. He argued that fair comparison could only be made when the densities involved were similar. He therefore attempted to delineate areas where minimum densities could be equated using the smallest administrative units. By calculating the population densities, he worked outward from the large cities until points were reached where the densities fell below a given figure. This allows a line to be drawn around the city so that all areas with a density over a given threshold were included (Gryztzell 1963).

Apart from the population density criterion, approaches using remote sensing technology have also been developed and applied for mapping land use, land cover, building density, climatic conditions, and socio-economic characteristics at various spatial and temporal scales in the urban environment (Phinn et al. 2002; Ward, Murray, and Phinn 2000; Hepner, et al. 1998; Mesev 1998; Barnsley and Barr 1997, 1996; Lo 1997; Harris and Ventura 1995; Mesev et al. 1995; Forster 1993, 1983; Gong and Howarth 1990; Moller-Jensen 1990). Remotely sensed data, especially those from the moderate resolution satellites such as Landsat and SPOT, offer highly useful sources for mapping the composition of cities and their changes over time; however, the use of remotely sensed data for mapping attributes of urban areas has met with limited success (Phinn et al. 2002). One of the most common limitations is the understanding of the boundary between an urban area and its non-urban hinterland as being a sharp line.

As cities are physical concentrations of population and housing, population or dwelling densities are measurable criteria in delimiting the extent of urban development. In particular, these criteria allow for the application of the fuzzy set theory to define a fuzzy concept of urban areas. In the following section, instead of defining a sharp boundary between urban and non-urban areas, a fuzzy set approach to delimiting the extent of urban development with population density is proposed. The criteria used in defining the membership functions are flexible, and they can be adjusted or calibrated according to different conditions when applied to individual cities.

3.1.3.2 Fuzzy Set Approach in Defining Urban Areas

Consider the fuzzy set "urban" defined earlier in Equation 3.2:

$$S_{urban} = \{(x_{ij}, \mu_{S_{urban}}(x_{ij})) | x_{ij} \in X\}$$

The value of the membership function $\mu_{S_{urban}}(x_{ij})$ ranges from 0 to 1, representing the state of a cell undergoing an urban development process. For instance, if a cell has a membership grade of 0, this cell has yet to be developed, that is, it is in a non-urban state; if a cell has a membership grade of 1, it has been fully developed as an urban area, and therefore this cell is in an urban state. Cells with a membership grade between 0 and 1 may have been developed to some extent, although they are yet to be fully developed.

In order to define a mathematical formula to represent the fuzzy membership function for delimiting urban areas, a population density criterion was employed as a measurable representation of the extent of urban development of cell x_{ij} in the urban fuzzy set S_{urban}. The population density value has been adjusted based on a number of other factors, such as dwelling density, major infrastructure such as sewerage and drainage supply, type of land use depicted from satellite images, and the percentage of population dependent on non-agricultural industries. For example, if an area has a low population density but the dwelling density is high, or if an area has a dominance of urban land-use type, then the population density needs to be adjusted to a higher value. In contrast, if an area has a relatively high population density but most of its land is used as farms, then the population density value needs to be reduced to some extent. Such factors were evaluated when processing raw data for Metropolitan Sydney.

A common approach for defining urban areas is evident from various countries using census data, that is, an area is regarded as urban if the population density of the area reaches a certain level. The assumption under this approach is that, if an area has a population density less than a certain value, this area is regarded as non-urban, and the grade of membership of this area in the urban fuzzy set is 0; if an area has a population density higher than another threshold value, it is regarded as a fully urban built-up area, and therefore its membership grade is 1. The lower and upper threshold values of population density can vary significantly from one country to another or even from one city to another. Therefore, these threshold values need to be defined individually according to situations in different countries or cities. For example, the lower threshold of population density excluding an area from being regarded as urban can be as low as 200 persons per square kilometre (appropriately 500 persons per square mile) in Australia, whereas this threshold in Canada and the United States should be 400 persons and 580 persons per square kilometre, respectively. With these threshold values, the membership function of areas in the urban fuzzy set can be defined as a function of its population density.

Let $\rho_{x_{ij}}$ be the population density of cell x at location i, j. The lower and upper thresholds of population density are represented by ρ_0 and ρ_1, respectively, in delimiting urban areas. To define the membership grade of cells in the urban fuzzy set, a simple linear membership function was employed, as shown in Equation 3.10:

$$\mu_{S_{urban}}(x_{ij}) = \begin{cases} 0 & \rho_{x_{ij}} < \rho_0 \\ \dfrac{\rho_{x_{ij}} - \rho_0}{\rho_1 - \rho_0} & \rho_0 \leq \rho_{x_{ij}} < \rho_1 \quad (x_{ij} \in X) \\ 1 & \rho_{x_{ij}} \geq \rho_1 \end{cases} \tag{3.10}$$

With this membership function, the population density values can be converted into the grade of membership. This involves matching the density measurement

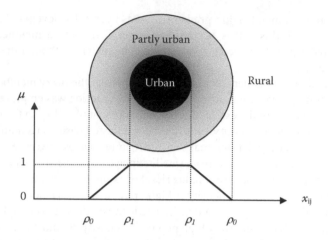

FIGURE 3.4 Defining an urban fuzzy set with a linear membership function. (μ is the membership function of cell x_{ij} in the urban fuzzy set; ρ_0 and ρ_1 are the lower and upper thresholds of population density, respectively, in delimiting urban areas.)

against the membership function (Figure 3.4), which is called a *fuzzification* process (Berenji 1992). Through this conversion, each cell in the urban fuzzy set receives one value of a membership grade, which represents the state of the cell in the urban fuzzy set. For instance, if the population density of a cell is less than the lower threshold, it receives a membership grade of 0. In this case, the state of the cell is regarded as "non-urban." If the population density of a cell is higher than the upper threshold, it receives a membership grade of 1, the state of which is regarded as "urban." All other cells receive a membership grade between 0 and 1, representing their extent of development in this urban fuzzy set. Their states in the urban fuzzy set are termed "partly urban." Therefore, instead of using a binary definition of non-urban and urban, multiple states in delimiting urban areas can be applied to simulate the continuous process of non-urban to urban conversion.

3.2 FUZZY LOGIC CONTROL IN CELLULAR AUTOMATA-BASED URBAN MODELLING

Urban development as a fuzzy process is not only represented in delimiting its urban extent, it is also represented in the factors driving such development. As a result of both physical constraints and human decision-making behaviour, urban development is a fuzzy logic-controlled process. This section introduces the methodology of fuzzy logic control in the simulation of urban development. As fuzzy logic control is developed based on two important concepts—the linguistic variable and fuzzy logic—these two concepts are discussed first, followed by discussions on the application of fuzzy logic control in urban modelling, especially in models using the cellular automata approach.

3.2.1 Linguistic Variables and Fuzzy Logic

3.2.1.1 Linguistic Variables

"In retreating from precision in the face of overpowering complexity, it is natural to explore the use of what might be called *linguistic* variables, that is, variables whose values are not numbers but words or statements in a natural or artificial language" (Zadeh 1975a: 201). An easy way to understand the notion of a linguistic variable is to regard it as a variable whose numerical values are fuzzy numbers, or as a variable the range of which is not defined by numerical values but by linguistic terms.

For example, consider a linguistic variable named "age"; the value of the variable ranges from 0 to 100 (years). However, instead of using numerical values, the value of this linguistic variable can be defined as linguistic terms such as "old," "young," "very old," "very young," and so on (Zadeh 1975a).

Let u be a variable representing the age in years of life. $\tilde{M}(x)$ is the rule that assigns a meaning—that is, a fuzzy set—to the linguistic terms. For instance,

$$\tilde{M}(\text{old}) = \{(u, \mu_{old}(u)) \mid u \in [0,100]\} \tag{3.11}$$

where μ_{old} is the membership function of u in the fuzzy set "old," which can be represented by the following formula (Zadeh 1975a):

$$\mu_{old}(u) = \begin{cases} 0 & u \in [0,50] \\ \left(1+\left(\dfrac{u-50}{5}\right)^{-2}\right)^{-1} & u \in (50,100] \end{cases} \tag{3.12}$$

In this example, a term set of the linguistic variable "Age" can be defined as Age = {Young, Very Young, Not Very Young, Not Very Old, Old, Very Old, Very Very Old, …}. An illustration of this linguistic variable is shown in Figure 3.5.

There are some special linguistic terms such as *very, more or less, fairly*, and *extremely* that can be used to modify the meaning of other linguistic terms. These are called linguistic hedges (or simply hedges or modifiers). Mathematical operations frequently used for modifiers include concentration, dilution, and contrast intensification. Using $\tilde{A}$ to represent a fuzzy set, these three operations can be expressed as

Concentration: $$\mu_{con(\tilde{A})}(u) = (\mu_{\tilde{A}}(u))^2 \tag{3.13}$$

Dilution: $$\mu_{dil(\tilde{A})}(u) = (\mu_{\tilde{A}}(u))^{1/2} \tag{3.14}$$

Contrast intensification: $$\mu_{int(\tilde{A})}(u) = \begin{cases} 2(\mu_{\tilde{A}}(u))^2 & \text{for } \mu_{\tilde{A}}(u) \in [0, 0.5] \\ 1-2(1-\mu_{\tilde{A}}(u))^2 & \text{otherwise} \end{cases}$$

$$\tag{3.15}$$

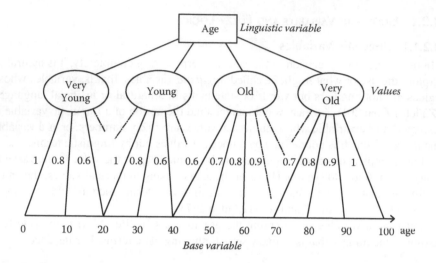

FIGURE 3.5 The linguistic variable "age." (Adapted from Zadeh, L. A. *Information Sciences* 8: 199–249. 1975. With permission.)

These mathematical operations can be associated with the following linguistic hedges (modifiers) as defined by Zimmermann (1987):

Very $\tilde{A}$ = con ($\tilde{A}$)
More or less $\tilde{A}$ = dil ($\tilde{A}$)
Plus $\tilde{A}$ = $(\tilde{A})^{1.25}$
Slightly $\tilde{A}$ = int [plus $\tilde{A}$ and not (very $\tilde{A}$)]

In addition, the following linguistic hedges are also used, which are measured at a 50% confidence level:

Nearly $\tilde{A}$ = $\tilde{A}$ ±5%
About $\tilde{A}$ = $\tilde{A}$ ±10%
Roughly $\tilde{A}$ = $\tilde{A}$ ±25%
Crudely $\tilde{A}$ = $\tilde{A}$ ±50%

For instance, if a number lies within 10% of 10, that is, the number is between 9 and 11, then we are at least 50% confident that the number belongs to the fuzzy set "about 10."

3.2.1.2 Basic Logic Terms and Reasoning

As a basis for reasoning, logic can be distinguished essentially by topic-neutral (context-independent) items: truth values, vocabulary (operators), and reasoning procedure (tautologies, syllogisms) (Zimmermann 1991). In Boolean logic, truth values can be 0 (false) or 1 (true), and by means of these truth values the vocabulary (operator) is defined via truth tables. For two statements with two truth-values of 0 or 1, there are 2^{2^2} = 16 truth tables, each defining an operator (Table 3.1).

TABLE 3.1

Truth Value Table for Two Statements with Two Truth Values

A	B	w_1	w_2	w_3	w_4	w_5	w_6	w_7	...	w_{16}
1	1	0	1	0	1	1	1	1	...	0
1	0	0	1	1	0	0	1	1	...	1
0	1	0	1	1	1	0	0	1	...	0
0	0	1	0	0	1	1	0	1	...	0

Note: A and B represent two statements, either of which can be true (shown as 1) or false (shown as 0). Various logic operations of the two statements are shown in w_1 to w_{16}.

Assigning meanings (words) to these operators is not difficult for the first four or five columns. For example, the first one (w_1) being the logic "and," the second one (w_2) "inclusive or," the third one (w_3) "exclusive or," and the fourth (w_4) and fifth (w_5) the "implication" and the "equivalence." However, assigning meanings to the last nine operations (w_7 to w_{16}) will be difficult in terms of the human language. If there are three statements rather than two, the task of assigning meanings to the truth tables becomes even more difficult (Zimmerman 1991).

The third topic-neutral item of a logical system—the reasoning procedure—is based on tautologies such as

Modus ponens: $(A \wedge (A \Rightarrow B)) \Rightarrow B$

Modus tollens: $((A \Rightarrow B \wedge \neg B)) \Rightarrow \neg A$

Syllogism: $((A \Rightarrow B) \wedge (B \Rightarrow C)) \Rightarrow (A \Rightarrow C)$

Contraposition: $(A \Rightarrow B) \Rightarrow (\neg B \Rightarrow \neg A)$

Here, A, B, and C represent three sets; symbols $\wedge$ and $\neg$ denote logical **AND** and logical NOT respectively; symbol $\Rightarrow$ shows a logical implication.

The Latin term *modus ponens*, meaning "mode that affirms by affirming," is a common form of reasoning that can be interpreted as "if A is true and if the statement 'if A is true then B is true' is also true, then B is true."

On the other hand, the Latin term *modus tollens* means "mode that affirms by denying," and it can be interpreted as "if the statement 'if A is true then B is true' is true and if B is false, then A is also false."

Syllogism is a Greek word that means "conclusion" or "inference." It can be interpreted in logic term as "if both the statements 'if A is true then B is true' and 'if B is true then C is true' are true, then the statement 'if A is true then C is true' is also true." Similarly, *contraposition* can be interpreted as "if the statement 'if A is true then B is true' is true, then the statement 'if B is false then A is false' is also true." In other words, for the statement "if A then B" for any two propositions A and B, the contraposition is "if not B then not A."

3.2.1.3 Fuzzy Logic

Fuzzy logic has been proposed by Zadeh (1975a,b,c) as an extension of classical formal models of reasoning into models that incorporate fuzziness; it is the application of the fuzzy set theory to human reasoning that is approximate rather than precisely deducted from classical Boolean logic. Fuzzy logic allows modelling complex systems using a higher level of abstraction originating from human knowledge and experience. This knowledge and experience can be expressed with subjective linguistic terms such as high, hot, or cold, which can be mapped into exact numeric ranges. According to Gaines (1976, 1975), there are at least three possible definitions for fuzzy logic.

1. A basis for reasoning with vague statements
2. A basis for reasoning with vague statements using the fuzzy set theory for the fuzzification for logic structures
3. A multivalued logic in which truth values are in the interval [0,1], and the valuation of a disjunction is the maximum of those of the disjuncts, and that of a conjunction is the minimum of those of the conjuncts

The first definition is a very general one, and the second a restricted form of the first—this comes close to the content in Zadeh's papers (1975a,b,c). The third definition, which is the widely used form, can be regarded as a set of infinitely valued logic that differs from others only in its implication functions (Kickert 1978).

The fundamental difference between classical logic and fuzzy logic is in the range of their truth-values. In fuzzy logic, the truth or falsity of fuzzy propositions is a matter of degree. Assuming that truth and falsity are expressed by values 1 and 0, respectively, the degree of truth of each fuzzy proposition is expressed by a number in the unit interval of [0, 1]. In fuzzy logic, the number of truth-values is, in general, infinite. Instead of using numbers, the truth-values are linguistic variables (or terms of the linguistic variable "truth"). The terms of the linguistic variable "truth" can be tabulated as a finite number of terms, such as true, very true, false, more or less false, very false, and so on. With this tabulation, the logic operators, such as "and," "or," and "not," are also defined in fuzzy logic, and the extensive principles can be applied to derive definitions of these operators.

Fuzzy logic is an organised and mathematical method of handling *inherently* imprecise concepts that classical Boolean logic cannot handle easily. For instance, there is no sharp cut-off between the concepts of "old" or "young," even though people generally have a concept of what an "old" or "young" person is. Rather, these concepts can be better represented using fuzzy set and fuzzy logic terms.

Consider the modus ponens in fuzzy reasoning. Let $\tilde{A}$, $\tilde{A}'$, $\tilde{B}$, $\tilde{B}'$ be four fuzzy statements; then the generalised modus ponens reads:

Premise:	x is $\tilde{A}'$
Implication:	if x is $\tilde{A}$ then y is $\tilde{B}$
Conclusion:	y is $\tilde{B}'$

For instance, if x represents an area, $\tilde{A}$ and $\tilde{A}'$ are fuzzy statements of "densely populated" and "very densely populated"; $\tilde{B}$ and $\tilde{B}'$ are fuzzy statements of "developed" and "highly developed," respectively, then the following reasoning can be achieved:

Premise: Area x is very densely populated ($\tilde{A}'$).

Implication: If an area is densely populated ($\tilde{A}$), then the area is developed ($\tilde{B}$).

Conclusion: Area x is highly developed ($\tilde{B}'$).

The fuzzy logic approach is a revolutionary technology in logic. It provides "a means of translating natural language-based expressions of knowledge and common sense into a precise mathematical formalism" (Openshaw and Openshaw 1997: 269). It also gives computers the ability to think and make decisions more like human beings (McNeill and Freiberger 1994). For geographers, it offers "a refreshingly new perspective on how to go about building better models of geographical systems by handling rather than ignoring or artificially removing the fuzziness within them" (Openshaw and Openshaw 1997: 269).

3.2.2 Fuzzy Logic Control

Fuzzy logic control is the application of the fuzzy set theory and fuzzy logic in control systems. It has been increasingly applied to solve problems in complex systems with embedded control and information processing. Instead of modelling a system mathematically, a fuzzy logic control system incorporates a simple, rule-based approach in the form of an "if–then" format to solve the control problem. This methodology provides a remarkably simple way to draw definite conclusions based upon vague, ambiguous, imprecise, noisy, or missing input information. The basic idea behind this approach is to incorporate the "experience" of a human process operator in the design of the controller. From a set of linguistic fuzzy control rules that describe the operator's control strategy, a control algorithm, in which the words are defined as fuzzy sets, is constructed. The main advantages of this approach are the possibility of implementing "rule of thumb" experience, intuition, and heuristics, and the fact that it does not need a model of the process (Kickert and Mamdani 1978). In a fuzzy logic control system, fuzzy logic is used to convert the heuristic control rules as stated by a human operator into an automatic control strategy (Mamdani and Assilian 1975).

In a fuzzy logic control system, the input variables are non-fuzzy measurable values. These input variables of the system will be transformed into "fuzzy" values, which are described in terms of membership in a fuzzy set. The first step in developing a fuzzy logic control system is called a *fuzzification process*. Typically, this involves setting up a universal space, defining linguistic variables, and selecting a membership function to convert the non-fuzzy input values into fuzzy terms. For example, Section 3.1.3.2 presented an approach using linguistic variables and a linear membership function to delimit the extent of urban areas based on the non-fuzzy values of population density. This is a fuzzification process, and it generates three fuzzy terms, namely, urban, partly urban, and non-urban. Similarly, a fuzzification process can also be developed to describe the slope of a landscape as flat, moderate, steep, and very steep.

The second process in developing a fuzzy logic control system is to develop a fuzzy information-processing engine with a set of rules or a rule base. This information-processing system is called a rule processor or a fuzzy inference engine. The fuzzy inference engine is developed to process information that is not perfectly clear and that reflects human knowledge, belief, and expertise. It is normally built as a computer software programme to employ inference steps similar to how the human brain works, including association, recognition, deduction, and decision making. Typically, the fuzzy inference rules appear as a set of simple "if–then" statements with fuzzy linguistic variables imbedded in the rules as modifiers.

For example, a simple "if–then" rule to control urban development of an area can be expressed as follows:

IF   the slope of an area is very steep,
THEN   development in this area should be very slow.

After the fuzzy inferencing, the last process of a fuzzy logic control system is the defuzzification process, which transforms results generated by the system into non-fuzzy quantifiable values and presents them as output of the system. This defuzzification process also involves setting up a universal space, and defining linguistic variables and an output membership function for quantification of the system's results. The whole process of a fuzzy logic control system is illustrated in Figure 3.6.

Fuzzy set and fuzzy logic control provide a linguistic non-numerical, non-mathematical, and non-statistical approach to modelling complex systems. This is a fairly simple approach with only a few rules being required to handle considerable complexity. It has found much practical applicability in industries and engineering systems (e.g., Umbers and King 1980; Kickert and Van Nanta Lemka 1976; Mamdani and Assilian 1975). It has also been applied in the social sciences, especially in simulating the human decision-making processes (see Kickert 1978; Wenstøp 1976).

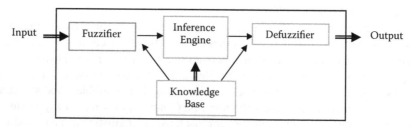

FIGURE 3.6 A fuzzy logic control system.

3.2.3 Fuzzy Logic Control in Cellular Automata-Based Urban Modelling

A cellular automaton is a dynamic system in which rules controlling the transition of its cells from one state to another can be deterministic, such as the rules applied in the "Game of Life" (Gardner 1972). However, the transition rules in a cellular automaton can also be non-deterministic, which are more appropriate when a cellular automaton represents a human-related system. For example, in an urban system, although it is understood that topography is a constraint to urban development, it is not correct to claim that "an area with a slope higher than 10 degrees cannot be developed into an urban area." Even though areas with smooth terrain are usually first selected for development, development could occur in areas with steep terrain under certain circumstances, such as a high demand for land but short supply of flat land, or people earning high incomes preferring to live on higher elevations to obtain good views of the natural landscape. Therefore, the terrain constraint is not a deterministic factor to urban development functioning on a "yes" or "no" basis.

Consider the rules in the model "Game of Life." The original rules were set up in deterministic ways. If these deterministic rules were changed to non-deterministic ones, such as

A live cell stays alive if it is *not crowded* in its neighbourhood; it dies either of *isolation* or *overcrowding*. A dead cell becomes alive if there are *sufficient* live cells in the neighbourhood.

By introducing a number of fuzzy linguistic variables such as "crowd," "not crowd," "isolate," "overcrowd," and "sufficient," the rules controlling the transition of cells from one state to another became fuzzy, no longer based on classical Boolean logic but on fuzzy logic. These fuzzy rules can be expressed as three "if–then" statements, each of which represents one rule of a cellular automaton.

Rule I:
    IF   it is *not crowded* in the neighbourhood of a live cell,
    THEN   this live cell stays alive in the next generation;
Rule II:
    IF   it is *isolating or overcrowding* in the neighbourhood of a live cell,
    THEN   this live cell dies in the next generation;
Rule III:
    IF   there are *sufficient* live cells in the neighbourhood of a dead cell,
    THEN   this dead cell becomes alive in the next generation.

Urban development is a complex spatial phenomenon controlled by many factors. The geographical conditions of the area, socio-economic status, infrastructure supply, demographic features, the potential of population growth, planning and zoning constraints, environmental protection regulations, as well as group and individual decision-making behaviours all play a role in the process of urban development. However, none of these factors functions in a deterministic manner. In other words, urban development is not controlled by Boolean logic. Instead, the controlling process of these factors is based on fuzzy logic. For instance, although it is not appropriate to say that "an area with a slope of 10 degrees or higher cannot be developed into an urban area," it will be true to say that "development is *less likely* to happen on land with a *steep* terrain." With the use of linguistic variables, the rule becomes fuzzy, and it functions according to the principles of fuzzy logic. Therefore, advantages would come by incorporating fuzzy logic control in a cellular automaton to simulate the process of urban development.

3.3 DEVELOPING FUZZY CONSTRAINED CELLULAR AUTOMATA FOR URBAN MODELLING

Previous sections argue that urban development is a fuzzy process that is controlled by many non-deterministic factors. This process is better represented as a fuzzy logic control system. In defining such a fuzzy logic control system, the state of urban extent needs to be fuzzified using a membership function to define the degree of membership in the urban fuzzy set. A set of rules need to be identified and a fuzzy inference engine constructed according to the nature of the system being modelled. By applying the rules and inference engine to the system, results will be generated that need to be defuzzified as quantifiable output of the system. This section first presents the temporal process of urban development based on classical research, followed by discussions on defining the speed of urban development using the fuzzy set concept. A number of fuzzy constrained transition rules and the fuzzy inference engine are proposed to model the process of urban development. The process of defuzzification of the fuzzy logic control system is also presented later in this section.

3.3.1 THE TEMPORAL PROCESS OF URBAN DEVELOPMENT

Section 3.1.3 described the spatial process of urban development. It identified a trend of continuous change from non-urban or rural to partly urban and then urban areas; there is no sharp boundary between non-urban and urban areas. Therefore, the fuzzy set theory is applicable in delimiting the fuzzy nature of the non-urban to urban transition.

Similar to its spatial process of development, urban development is also a continuous process over time. Generally, this temporal process of development follows a logistic curve (Herbert and Thomas 1997; Jakobson and Prakash 1971; Fourastié 1963). Fourastié (1963) suggests that the period of industrialisation and urbanisation is a transitory stage in the history of mankind, during which societies transform from primary or an agriculture-based stage to tertiary or a service-occupation-based stage. The transition stage is divided into three parts labelled take-off, expansion, and achievement. The process of tertiary civilisation follows a logistic curve, progressing from 10 to 80% levels of urbanisation in a society (Figure 3.7). This logistic trend of development has been identified in many places around the world (Figure 3.8).

To interpret the logistic curve of urban development verbally, when a region starts to be urbanised through the development process, it is normally quite slow in the initial stage (the take-off part). After the initial stage of slow development, it will speed up in the next stage (the expansion part) until it gets to a high standard of urban status. Then development will slow down until the region is fully developed (the achievement part).

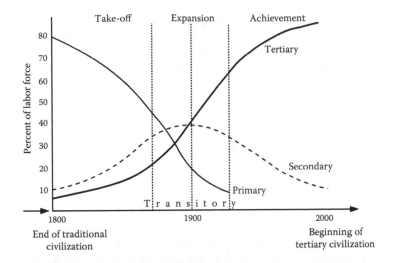

FIGURE 3.7 Urban development and the progression from transitional to modern. (Adapted from Herbert, D. T. and Thomas, C. J., *Cities in space: City as place,* 3rd ed., David Fulton Publishers, London. 1997. With permission.)

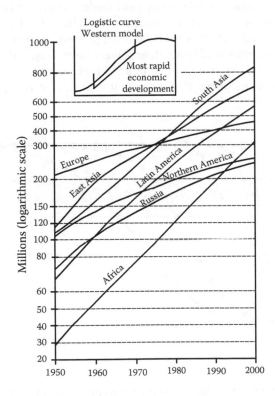

FIGURE 3.8 Logistic curve of urban development between 1950 and 2000. Y axis is population in millions; X axis is time. (Adapted from Herbert, D. T. and Thomas, C. J., *Cities in space: City as place,* 3rd ed., David Fulton Publishers, London. 1997. With permission.)

Assume that it takes n years for a non-urban region to be fully urbanised. Based on the understanding of the logistic curve of the urban development process, the relationship between the state of the cell in the urban fuzzy set and the year of development can be expressed in a mathematical term as

$$\mu_{S_{urban}}\left(x_{ij}^t\right) = \begin{cases} 0 & t_{x_{ij}} = 0 \\ \dfrac{1}{a + b \cdot \exp(-c \cdot t_{x_{ij}})} & 0 < t_{x_{ij}} < n \\ 1 & t_{x_{ij}} = n \end{cases} \qquad (3.16)$$

where $\mu_{S_{urban}}(x_{ij}^t)$ is the membership function of the cell x_{ij} at time t in the urban fuzzy set; $t_{x_{ij}}$ denotes the t-th year of development of the cell x_{ij} in the urban development process; and a, b, and c are parameters of the logistic function that define the shape of the logistic curve.

Studies of the logistic function show that the shape of the logistic curve is not overly sensitive to changes made through parameters a and b; however, it is very sensitive to changes made through parameter c (Figure 3.9). Moreover, the parameter c is related to the number of years for a full process of urban development. For instance, if the full process of urban development takes longer, the value of c will be smaller; otherwise, its value will be larger. Pragmatically, the parameter c can be defined as $c = \psi/n$, where ψ is a constant value and n is the total number of

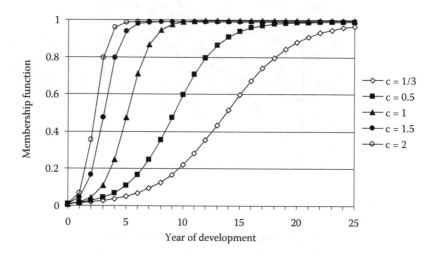

FIGURE 3.9 Sensitivity of the logistic function to its parameter. (This figure shows that the membership function of the urban fuzzy set is very sensitive to changes made through parameter c of the logistic function.)

years for a non-urban region to be fully urbanised. Therefore, Equation 3.16 can be rewritten as

$$
\mu_{S_{urban}}\left(x_{ij}^{t}\right) =
\begin{cases}
0 & t_{x_{ij}} = 0 \\[2ex]
\dfrac{1}{a + b \cdot \exp(-\psi \cdot t_{x_{ij}} / n)} & 0 < t_{x_{ij}} < n \\[2ex]
1 & t_{x_{ij}} = n
\end{cases}
\tag{3.17}
$$

The parameters a, b, ψ, and n in Equation 3.17 need to be configured and calibrated to fit the actual urban development process of individual cities.

Equation 3.17 can also be rewritten as Equation 3.18 to compute the temporal stage of a cell in the urban development process, based on the state of the cell in the urban fuzzy set.

$$
t_{x_{ij}} =
\begin{cases}
0 & \mu_{S_{urban}}\left(x_{ij}^{t}\right) = 0 \\[2ex]
\left[\ln(b) - \ln\left(\dfrac{1}{\mu_{S_{urban}}\left(x_{ij}^{t}\right)} - a \right) \right] \cdot n / \psi & 0 < \mu_{S_{urban}}\left(x_{ij}^{t}\right) < 1 \\[2ex]
n & \mu_{S_{urban}}\left(x_{ij}^{t}\right) = 1
\end{cases}
\tag{3.18}
$$

For each cell x_{ij}, its membership value $\mu_{S_{urban}}(x_{ij})$ in the urban fuzzy set can be computed using a membership function, such as the one shown in Equation 3.10. Therefore, with the awareness of the current state of a cell in the urban development process, the stage (or year as denoted by $t_{x_{ij}}$) of development of the cell in the urban development process can be identified by Equation 3.18.

In addition to continued urban development from partly developed areas to a fully urban state, new development can also occur in some undeveloped areas with advantageous conditions for development. For instance, if a non-urban area is surrounded by areas that are highly urbanised, this non-urban area may have the tendency to be influenced by its neighbouring areas to develop toward an urban state. In another situation, with the construction of a new railway line extending from an urban to a rural region, some areas along the new railway line could be selected for urban development. In this case, an initial low membership grade may be assigned to the non-urban cell so that it will start developing toward a partly urban, and then urban, state.

3.3.2 THE SPEED OF URBAN DEVELOPMENT AS A FUZZY SET

Obviously, the number of years for a cell to be fully developed into an urban state is associated with the speed of development, which can be fast or slow, depending on a number of controlling factors. These include the conditions of the cell itself, geographic conditions of the areas, socio-economic status, as well as planning and government policy controls, resulting in variations in the pattern and speed of development in space and time. For instance, in a fast growing area with sufficient infrastructure and transportation support, development will be faster compared to an area without sufficient infrastructure supply and transportation support. The speed of development varies from one region to another and from one locality to another within the same region.

Therefore, the *speed* of development is a fuzzy term that can be divided into a range of "states" modified by some linguistic variables, such as "very fast," "fast," "medium," "slow," or "very slow." This results in a range of n values in Equation 3.17. For instance, a medium process of urban development may take around 20 years, whereas a fast process may take 10 to 15 years. Similarly, a slow process of urban development may take 20 to 25 years, or even longer if the process is very slow.

Assume that the full process of urban development ranges from 1 to 30 years or longer. In order to reflect the gradual change of urban development from one speed to the next, fuzzy membership functions are used to represent the magnitude of the speed in each state. The membership function can take a simple triangular shape, as is shown in Figure 3.10.

Using η_{ef}, η_{vf}, η_f, η_m, η_s, η_{vs}, and η_{es} to represent the membership functions for "extremely fast," "very fast," "fast," "medium," "slow," "very slow," and "extremely slow," respectively, these seven linguistic variables can be defined as follows:

Extremely fast: If the whole process of urban development is completed within 5 years, this development is extremely fast. Therefore, the speed of urban development will gain full membership in the "extremely fast" state. However, if the process takes more than 5 years but less than 10 years, it will still gain partial membership in the "extremely fast" state. Therefore, the membership function for "extremely fast" can be expressed as

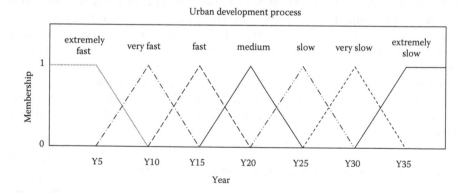

FIGURE 3.10 Membership functions defining the "speed" of urban development.

$$\eta_{ef}\left(x_{ij}^t\right) = \begin{cases} 1 & n \le 5 \\ 2-0.2n & 5 < n \le 10 \end{cases} \tag{3.19}$$

Very fast and Fast: Similar to the definition for "extremely fast," the linguistic variable "very fast" and "fast" ranges from 5 to 15 years and 10 to 20 years, respectively, with a full membership of 1 in the 10th and 15th year of development for "very fast" and "fast," respectively.

$$\eta_{vf}\left(x_{ij}^t\right) = \begin{cases} -1+0.2n & 5 < n \le 10 \\ 3-0.2n & 10 < n \le 15 \end{cases} \tag{3.20}$$

$$\eta_{f}\left(x_{ij}^t\right) = \begin{cases} -2+0.2n & 10 < n \le 15 \\ 4-0.2n & 15 < n \le 20 \end{cases} \tag{3.21}$$

Medium: The linguistic variable "medium" ranges from 15 to 25 years, with a full membership of 1 in the 20th year of development.

$$\eta_{m}\left(x_{ij}^t\right) = \begin{cases} -3+0.2n & 15 < n \le 20 \\ 5-0.2n & 20 < n \le 25 \end{cases} \tag{3.22}$$

Slow and Very slow: The linguistic variable "slow" ranges from 20 to 30 years, with a full membership of 1 in the 25th year of development. Similarly, the linguistic variable "very slow" ranges from 25 to 35 years, with a full membership of 1 in the 30th year of development.

$$\eta_{s}\left(x_{ij}^t\right) = \begin{cases} -4+0.2n & 20 < n \le 25 \\ 6-0.2n & 25 < n \le 30 \end{cases} \tag{3.23}$$

$$\eta_{vs}\left(x_{ij}^t\right) = \begin{cases} -5+0.2n & 25 < n \le 30 \\ 7-0.2n & 30 < n \le 35 \end{cases} \tag{3.24}$$

Extremely slow: Urban development is considered "extremely slow" if it takes more than 30 years to develop; full membership in this "extremely slow" state starts when it takes more than 35 years to be fully developed. Therefore,

$$\eta_{es}\left(x_{ij}^t\right) = \begin{cases} -6+0.2n & 30 < n \le 35 \\ 1 & n > 35 \end{cases} \tag{3.25}$$

Of course, different experts would have different feelings about what speed should be classified as fast, medium, or slow, and so on. Therefore, leaving a good deal of overlap on the adjacent membership functions lends robustness to the reasoning system. With their membership functions, the speed of development will not jump abruptly from "very fast" to "fast," or from "medium" to "slow." Instead, as the speed changes, it loses membership value in one state and gains value in the next state. For instance, if the full process of urban development takes 15 years, it will have a full membership value of 1 in the "fast" speed; however, if it takes 12 years for the area to be fully urbanised, then the speed will have a membership value of 0.4 in the "fast" state, and 0.6 in the "very fast" state.

3.3.3 THE FUZZY TRANSITION RULES AND INFERENCING

In order to define transition rules for urban development, it is assumed that the process of urban development is non-reversible, that is, it develops from non-urban to partly urban and then urban; there is no anti-urbanisation process. Once a cell is fully urbanised, it will stay in the urban state; no urban re-development or consolidation is considered in this model. With these assumptions, the transition matrix from non-urban to partly urban and then urban is demonstrated in Table 3.2.

3.3.3.1 Primary Transition Rules

According to the principles of cellular automata theory, the state of cells in an urban fuzzy set at a certain time is determined by the state of the cell itself in a previous time step and the state of the cells of its neighbourhood through a set of transition rules. This is the fundamental transition rule that can be applied to model the development of an urban system based on the fuzzy set concept.

> If a cell has the propensity for development and it can receive sufficient support for development from its neighbourhood, this cell will develop at a medium speed to a higher state in the urban fuzzy set.

However, the development will be slowed down if its own propensity for development is weak, or if it cannot receive sufficient support for development from its neighbourhood. On the other hand, this development can also be speeded up if the cell has a strong propensity for development and/or it receives strong support for development from its neighbourhood.

TABLE 3.2

Urban Transitions from Non-Urban to Partly Urban and Urban

T2 ╲ T1	Non-urban	Partly urban	Urban
Non-urban	No change in state and membership	Not allowed	Not allowed
Partly urban	Change in state and membership	No change in state; with or without change in membership	Not allowed
Urban	Change in state and membership	Change in state and membership	No change in state and membership

Note: This table shows the possible transition of cell states from one time (T1) to another (T2).

Using a series of "if–then" statements, the following rules are proposed as the primary controller to the transition of cells from one state to another.

Primary Rule 1 (PR1):
    IF   a cell is in a *partly* urban state with *strong* propensity to further development,
    AND   the cells in the neighbourhood are developed to a *higher* state than the cell in question,
    THEN   the cell will develop at a *medium* speed.
Primary Rule 2 (PR2):
    IF   a cell is in a *partly* urban state with *strong* propensity to further development,
    AND   the cells in the neighbourhood are also *partly* urbanised, but their states are *lower* than the cell in question,
    THEN   the cell will develop at a *slow* speed.
Primary Rule 3 (PR3):
    IF   a cell is in a *partly* urban state with *weak* propensity to further development,
    AND   the cells in the neighbourhood are developed to a *higher* state than the cell in question,
    THEN   the cell will develop at a *slow* speed.
Primary Rule 4 (PR4):
    IF   a cell is in a *partly* urban state with *weak* propensity to further development,
    AND   the cells in the neighbourhood are *rarely* urbanised,
    THEN   the cell will develop at a *very slow* speed.
Primary Rule 5 (PR5):
    IF   a cell is in a *non-urban* state,
    AND   the cells in the neighbourhood are *highly* urbanised,
    THEN   the cell may start to develop at a *slow* speed.
Primary Rule 6 (PR6):
    IF   a cell is in a *non-urban* state,
    AND   the cells in the neighbourhood are *partly* urbanised to some extent but yet to be fully urbanised,
    THEN   the cell may develop at a *very slow* speed.

3.3.3.2 Rule Firing Threshold

All rules are in abstract status when they were first constructed and subsequently entered into a computer program; they cannot do anything but to wait for the right data to be entered. If a rule is turned on and the input data satisfies the antecedent, that is, the "IF" requirements of the rule, including the AND part, then the rule will fire, and the system will take the consequent action of the rule as specified in the "THEN" statement.

In a fuzzy rule-based system, it is not completely clear whether the input data matches the antecedent of a rule. For instance, if a cell has a membership value of 0.7 in the urban fuzzy set, it means that the cell is 70% urbanised, and therefore the cell is considered to have the propensity to be further developed. If the memberships of its neighbouring cells are of a value of 0.8 or higher in the urban fuzzy set, this means that the cell will also receive support from its neighbourhood to be further developed. In this case, PR1 will fire. However, what if a cell has a membership value of around 0.5 and the membership values of its neighbouring cells are also around 0.5? In this case, a minimum confidence level needs to be set up to accept the input data as a match for the antecedent of the rule. This minimum confidence level is called *rule-firing threshold*. If the input data generates a confidence level equal to or higher than the rule-firing threshold, the rule will fire, and the consequent actions that are specified in the rule will take place.

Using x_{ij} and $\mu_{S_{urban}}$ to represent the cell in question and its membership grade in the urban fuzzy set, respectively, μ_0 is used to represent the rule-firing threshold, or the minimum membership value for a cell to have strong propensity for further development. The value of μ_0 can be initially set to 50% or 0.5 membership in the urban fuzzy set; this value can be adjusted during the model calibration process.

If cells within the neighbourhood of cell x_{ij} are represented by $\Omega_{x_{ij}}$, $\mu_{\max S_{urban}}(\Omega_{x_{ij}})$ is used to represent the largest membership value of cells within the specified neighbourhood. If a cell can receive support for development from its neighbourhood, this means that the membership grade of some or all of the neighbouring cells should be larger than the membership grade of the cell itself. Hence, the six primary rules proposed in the previous section can be represented using their membership values as follows:

PR1:

IF $\qquad \mu_{S_{urban}}(x_{ij}) \geq \mu_0$ AND $\mu_{\max S_{urban}}(\Omega_{x_{ij}}) > \mu_{S_{urban}}(x_{ij})$

THEN $\qquad$ development will occur to the cell at a *medium* speed.

PR2, PR3:

IF $\qquad \mu_{S_{urban}}(x_{ij}) \geq \mu_0$ AND $\mu_0 \leq \mu_{\max S_{urban}}(\Omega_{x_{ij}}) < \mu_{S_{urban}}(x_{ij})$, OR

$\qquad\qquad 0 < \mu_{S_{urban}}(x_{ij}) < \mu_0$ AND $\mu_{\max S_{urban}}(\Omega_{x_{ij}}) > \mu_{S_{urban}}(x_{ij})$

THEN $\qquad$ development will occur to the cell at a *slow* speed.

PR4:

IF $\qquad 0 < \mu_{S_{urban}}(x_{ij}) \leq \mu_0$ AND $0 < \mu_{\max S_{urban}}(\Omega_{x_{ij}}) < \mu_{S_{urban}}(x_{ij})$

THEN $\qquad$ development will occur to the cell at a *very slow* speed.

PR5:

IF $\qquad \mu_{S_{urban}}(x_{ij}) = 0$ AND $\mu_{\max S_{urban}}(\Omega_{x_{ij}}) \approx 1$

THEN $\qquad$ development may occur to the cell at a *slow* speed.

PR6:

IF $\qquad \mu_{S_{urban}}(x_{ij}) = 0$ AND $\mu_0 \leq \mu_{\max S_{urban}}(\Omega_{x_{ij}}) < 1$

THEN $\qquad$ development may occur to the cell at a *very slow* speed.

Apart from these transition rules, if a cell does not have the propensity for development, and neither can it obtain support for such development from its neighbourhood, then no development can occur to the cell; the state of the cell stays unchanged.

3.3.3.3 Secondary Transition Rules

The primary transition rules deal with an ideal situation where development is only controlled by the state of the cell itself and the state of cells of its neighbourhood. This situation only exists in von Thünen's "Isolated State" (von Thünen 1826). In practice, urban development is affected by many other factors such as topographic conditions, transportation networks, socio-economic status, as well as planning and human decision-making behaviours. Some factors, such as the provision of urban infrastructure and transportation network, will function as accelerators to speed up the process of urban development; others, such as a mountainous landscape, or an area with a lack of urban infrastructure supply, may function as constraints to slow down the development process. These accelerators or constraints should also be built into the transition rules of the model.

The accelerating or constraining factors on urban development can be implemented as secondary rules to modify the primary rules on the transition of cell states. The general principles on the inferencing of transition rules are that, if there is an accelerating factor within the neighbourhood of the cell in question, the speed of urban development of that cell will be upgraded one step higher in the "speed" fuzzy set; for instance, from medium to fast or from very slow to slow speed. On the other hand, if there is a constraining factor within the neighbourhood of the cell in question, the speed of urban development of the cell will be downgraded one step lower in the "speed" fuzzy set; for instance, from medium to slow, or from slow to very slow speed. If there is more than one such factor, the speed will be upgraded or downgraded two steps up or down. On the other hand, the existence of an accelerator and a constraint will cancel the effect of both factors; hence, the speed of development will remain unchanged.

With this inferencing engine, the following rules can be proposed to demonstrate the effect of some accelerating or constraining factors.

Secondary Rule 1 (SR1):
> IF   a cell is developing at a *medium* speed,
> AND   one accelerator exists within the neighbourhood of the cell,
> THEN   the cell will develop at a *fast* speed.

Secondary Rule 2 (SR2):
> IF   a cell is developing at a *medium* speed,
> AND   more than one accelerator exists within the neighbourhood of the cell,
> THEN   the cell will develop at a *very fast* speed.

Secondary Rule 3 (SR3):
> IF   a cell is developing at a *medium* speed,
> AND   one constraint exists within the neighbourhood of the cell,
> THEN   the cell will develop at a *slow* speed.

Secondary Rule 4 (SR4):
> IF   a cell is developing at a *medium* speed,
> AND   more than one constraint exists within the neighbourhood of the cell,
> THEN   the cell will develop at a *very slow* speed.

Secondary Rule 5 (SR5):
    IF   a cell is developing at a *medium* speed,
    AND   one constraint and one accelerator exist within the neighbour-
             hood of the cell,
    THEN   the cell will still develop at a *medium* speed.

Secondary Rule 6 (SR6):
    IF   a cell is in a non-urban state,
    AND   one accelerating factor exists within the neighbourhood of
             the cell,
    THEN   the cell may start to develop at a *slow* speed.

For new development of cells from non-urban state, once a development is initiated, further development of the cell is considered as continued development, the speed of which will be determined by the primary and secondary transition rules for continued development.

It should be noted that the accelerating and constraining factors for urban development may also be fuzzy variables. That is, such an accelerating or constraining factor can be strong, medium, or weak in affecting urban development. For instance, when introducing the topographic condition of an area into the model as a constraint on urban development, the land slope can be very steep, which will function as a strong constraint; or the slope can also be medium or not so steep, having a medium degree constraining impact on the development; or it may be flat, with no constraining impact. Therefore, while incorporating the impact of any accelerating or constraining factors into the fuzzy logic controller, the degree or extent of such impact on urban development also needs to be considered and built into the controller. The different impact of such factors can be reflected in the range of the speed fuzzy set.

For instance, it may range from 15 to 25 years for development to proceed at a medium speed. A cell will gain full membership in the "medium" speed fuzzy set if it takes 20 years for the cell to be fully developed. If there is an accelerator within the neighbourhood of the cell, development will be accelerated to a fast speed; therefore, the full process of development will take 10 to 20 years to complete. However, with a strong accelerator, the speed of development tends to be closer to the lower range, that is, closer to 10 years in the "fast" speed fuzzy set. If the accelerator is weak, the speed of development tends to be closer to the higher end, that is, closer to 20 years; and if the accelerator is of a medium strength, the speed will tend to stay in the middle, that is, around 15 years.

A constraining factor may have similar impacts on the process of urban development depending on its strength. For instance, if a constraining factor with medium strength pulls the speed of development from medium to slow (i.e., 20 to 30 years), then a strong constraint tends to pull the speed to the higher end of the "slow speed" fuzzy set, that is, closer to 30 years; however, if the constraint is weak, the speed of development may stay in the lower end of "the slow speed" fuzzy set, that is, closer to 20 years. Table 3.3 provides two exemplar transition rules, one incorporating the impact of land slope as a constraint on

TABLE 3.3

Transition Rules and Fuzzy Inferencing

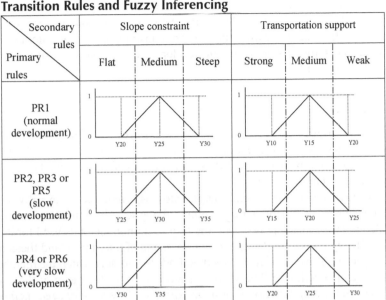

Secondary rules \ Primary rules	Slope constraint			Transportation support		
	Flat	Medium	Steep	Strong	Medium	Weak
PR1 (normal development)						
PR2, PR3 or PR5 (slow development)						
PR4 or PR6 (very slow development)						

urban development, and the other incorporating transportation network as an accelerator on such development.

With these fuzzy inferencing mechanisms, more factors can be introduced into the fuzzy logic controller individually or in combination with one another through the above procedure, depending on our understanding of the urban system and the availability of data that reflect factors affecting urban development. For a particular cell in the fuzzy logic control system, the question of which transition rule fires at a particular time depends on the condition of the cell itself and the conditions of cells in its neighbourhood. These conditions can be either physical, socio-economic, institutional, or a combination of any or all.

3.3.3.4 Rule Calibration

Obviously, the fuzzy inferencing rules need to be calibrated before they can be applied to simulate the actual process of urban development. For instance, PR1 shows that, under no accelerating or constraining conditions, a partly urbanised cell with a strong propensity for development will undergo a medium speed of development if it can get support for development from its neighbourhood. However, the question of how long this urban development can stay at a *medium* speed needs to be addressed; whether a *fast* or *slow* rather than *medium* speed can be applied to better reflect its real development process also needs to be evaluated and calibrated.

The calibrating of the control system can be a process of "trial and error." This process involves running the model to evaluate the results of the system; tuning the membership functions, rules, as well as the fuzzy inferencing mechanisms based on the feedback from the system; and retesting the system until satisfactory results are obtained. A full process of calibrating a fuzzy constrained cellular automata model using data illustrating historical urban development in Sydney, Australia, will be presented in Chapter 5.

3.3.4 THE DEFUZZIFICATION PROCESS

Once the fuzzy transition rules are applied, the results of the fuzzy logic controller are a set of fuzzy values representing the state of cells at a specific time spot. These fuzzy values need to be defuzzified into crisp values before they can be sent out as results of the system for action.

In order to defuzzify the system's output into crisp values, it is generally required to define membership functions for the output fuzzy set. There are many techniques available for defuzzification; however, the three common methods used for defuzzification of fuzzy output are the max criterion, mean of maximum, and the centre of area approaches. The *max criterion* method finds the point at which the membership function is at its maximum. The *mean of maximum* takes the mean of those points where the membership function is at its maximum. The *centre of area* method calculates the centre of gravity of the output fuzzy sets (Lee 1990).

However, no systematic justification is available for choosing a defuzzification strategy (Lee 1990). It is the human intuition that defuzzification is the reverse process of fuzzification. Therefore, the two processes should be reversible. For instance, if a number is fuzzified into a fuzzy set and it is immediately defuzzified, the same number should be achieved. Obviously, the process of defuzzifying a fuzzy set requires knowing representative values that correspond to each output fuzzy set member. Such representative values are related to the membership function used to defuzzify the input variables.

In the case of urban modelling developed in this book, a simple linear membership function is proposed to fuzzify the state of cells with a membership grade ranging from 0 (non-urban state) to 1 (fully urban state), representing an indefinite cell state in the urban fuzzy set. By applying the fuzzy transition rules, the state of cells of the model's outcomes also ranges from 0 to 1.

A simple linear membership function is applied in defuzzifying the system's output results; the function is similar to the one used in fuzzifying the input values of the cell states (Figure 3.11). With this membership function, the state of cells can be reversed back into three categories: non-urban, partly urban, and urban. Cells with a membership grade of 0 were defuzzified as non-urban, cells with a membership grade of 1 were urban, and all other cells whose membership grade ranged between 0 and 1 exclusively were categorised as partly urban cells.

It should be noted that, although the process of defuzzification is necessary to generate crisp output results for action, much information is lost by doing this, and further work needs to be done on how to use the information available in the solution fuzzy set (John 1995).

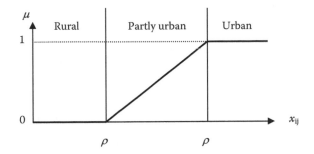

FIGURE 3.11 Membership function for defuzzifying the fuzzy output results.

3.4 CONCLUSION

This chapter developed a generic model of urban development based on the principles of the cellular automata and incorporated fuzzy set and fuzzy logic approaches. Compared with other cellular automata-based urban models, this model possesses advantageous features for simulating an urban development process. One of these features is the definition of cell states. Based on the fuzzy set theory, the state of a cell is associated with a grade of membership representing the stage of a cell in its urban development process. The grade of membership represents urban development as a continuous process in space and time, rather than as a binary transition from non-urban to urban.

Moreover, by constructing a fuzzy logic controller, the rules controlling the speed of transition of cells from one state to another are defined by a series of "IF–THEN" statements. These "IF–THEN" statements are presented using various linguistic variables, such as "quick," "very quick," "medium," "slow," "very slow," and so forth. Factors affecting the speed of urban development can also be implemented in the model as accelerators or constraints. These factors include physical environmental conditions, socio-economic status, as well as planning or other institutional constraints. Such factors are incorporated in the model as secondary rules to affect the process of urban development. The use of the fuzzy logic controller in defining transition rules makes the modelling of urban development closer to the human decision-making process. Before the model can be applied to simulate a real urban development process, it needs to be rigorously tested to evaluate and validate its appropriateness and suitability.

4 Sydney
Urban Development and Visualisation

As the capital city of the state of New South Wales (NSW) in Australia, Sydney, in the past 220 years, has developed from a penal settlement to one of the world's greatest cities with distinctive characters. The harbour, the beach, the rivers, the mountains, the home gardens and big backyards, and the large tracts of natural bushland make it unique among world cities (Spearritt and DeMarco 1988). Understanding how this city has been growing and foreseeing the future direction of its growth are of primary concern to geographers as well as to planners and decision makers.

In this and the following chapters, the fuzzy constrained cellular automata model of urban development constructed in Chapter 3 is applied to simulate the urban development of Sydney from 1976 to 2031. This chapter presents a description of the study area, its physical settings, the historical threads of urban development, as well as the impact of institutional factors on the development of the city.

The collection of geographical data illustrating various aspects of urban development and the processing of data in a geographical information system (GIS) are demonstrated in Section 4.2. These data sets are important in feeding the fuzzy constrained cellular automata model of urban development. Section 4.3 defines Sydney's urban areas with census data using the fuzzy set approach. The spatial patterns and temporal process of the urban development of Sydney from 1976 to 2006 are visualised in GIS. Conclusions on both data collection and the urban development of Sydney from 1976 to 2006 are addressed in Section 4.4, which also sets up the objectives of modelling the urban development of Sydney for the following chapters.

4.1 SYDNEY'S URBAN DEVELOPMENT AND PLANNING

Sydney is located on the south-east coast of Australia. It occupies an area between Broken Bay in the north and Port Hackings in the south. However, the concept of Metropolitan Sydney has been changing over time. Prior to the 1960s, Metropolitan Sydney was defined within the County of Cumberland Region (CCR). This is an area bounded by the Nepean–Hawkesbury River and its tributary. This area was expanded to include the whole Sydney Statistical Division defined by the Australian Bureau of Statistics (ABS), which was used in the Sydney Region Outline Plan (New South Wales State Planning Authority 1968). From the 1990s, a broad concept of the Greater Metropolitan Region—Sydney, together with Newcastle, the Central Coast, and Wollongong along the east coast of NSW—was identified in the planning scheme of the state government (New South Wales Department of Planning 1995).

However, the most recent release of the Sydney Metropolitan Strategy in 2005 defined a much more focused Sydney Metropolitan Area, which largely falls within the CCR (New South Wales Department of Planning 2005). Figure 4.1 shows the various spatial concepts of Metropolitan Sydney.

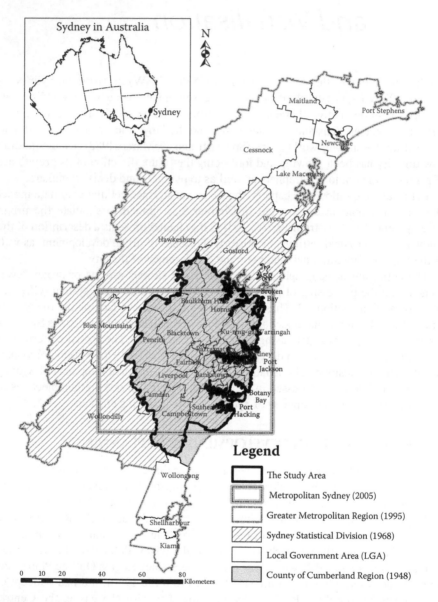

FIGURE 4.1 The different spatial concepts of Metropolitan Sydney. (Data sources: Census products 2006, Australian Bureau of Statistics; County of Cumberland Council 1948; New South Wales State Planning Authority 1968; New South Wales Department of Planning 1995.)

The study area selected to apply the cellular automata model of urban growth primarily follows the boundaries of the CCR, except for the south-east corner, which is geographically collected to the Illawarra Region in NSW and is outside the Sydney Statistical Division (Figure 4.1). This study area covers a land area of 3935 km². As most of the areas outside the CCR are extensive national parks having mountains and bushland, urban expansion outside its boundary is largely restricted by these natural barriers (Figure 4.2).

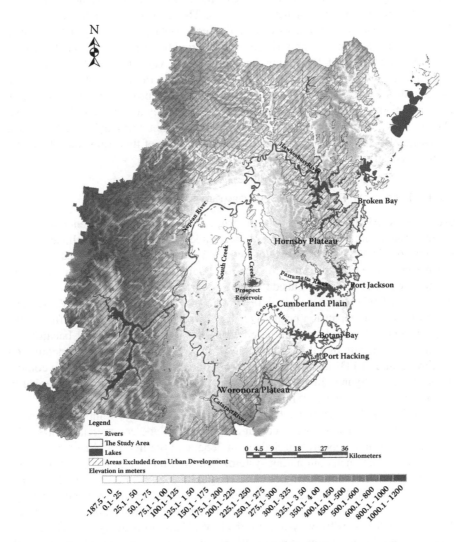

FIGURE 4.2 Geographic conditions of Sydney and its surrounding areas. The data layer "Areas Excluded from Urban Development" is defined to include water bodies (including lakes, reservoirs, and water courses), water supply reserves, forestry reserves, nature conservation reserves, prohibited areas for defence purposes, and mines or cemetery areas where urban development is strictly constrained. (Data source: Geoscience Australia 2006.)

Geographically, the CCR consists of three parts: the Hornsby Plateau to the north, the Woronora Plateau to the south, and the Cumberland Plain in between. The Hornsby and Woronora sandstone plateaus are quite similar in geology. Both plateaus contain hard Hawkesbury sandstone with small shale layers, which produced poor soils but good building material (Haworth 2003). The topographies rise to elevations of 270 m and 440 m, respectively, and are divided into a series of ridges and gorges with smaller branching ridges and steep gullies, which has forced the direction of set-tlement along the flatter interfluves. Some of the areas are defined as national parks, national recreational areas, or nature reserves. The Cumberland Plain lies between the two sandstone plateaus and extends to the sea at Botany Bay. The geology type of the Cumberland Plain is mainly shales, which provides fair grazing soils (Haworth 2003). For the most part, the Cumberland Plain is flat with a relief of less than 50 m, or a gently undulating terrain through which wind the Parramatta and Georges Rivers and a number of creeks flowing north and west into the Nepean River (Winston 1957). The Parramatta River passes through the centre of Sydney to the sea at Port Jackson. The Georges River comes together from several little tributaries and flows from west to east into Botany Bay. The topographic characteristics of this region have been the strongest constraints on the city's growth (Haworth 2003).

4.1.1 HISTORICAL THREADS OF DEVELOPMENT

Sydney was originally occupied by aboriginal people for many generations. Its urban development began when the first fleet under Captain Arthur Phillip arrived at Sydney Cove in 1788; but the early growth of Sydney was very slow because of the despotism of its early governors, the unwilling labour of its convict population, and its utter dependence on the outside world (Spearritt 2000; Spearritt and DeMarco 1988; Aplin 1982). It was not until 1810, with the arrival of Governor Macquarie, that Sydney's real expansion began. During Governor Macquarie's term from 1810 to 1821, Sydney's population increased from 11,950 to 38,778, and it became the administrative centre of a flourishing colony. In 1823, free settlers were encouraged to migrate, and in 1831 assisted migration commenced. When the last of the convict ships arrived in Sydney in 1849, vast areas of Sydney were available for people to live in (Frost and Dingle 1995).

From 1851 to 1861, with the discovery of gold, a very large number of migrants came to Sydney, transforming it into an administrative and commercial centre. The first railway system appeared in Sydney in 1855, and the first tramline was opened in 1861. However, such commuting systems had little impact on the urban develop-ment of Sydney in the first two decades (Fitzgerald 1987; Roberts 1978). It was not until the 1880s and 1890s that the construction of suburban railways and tramways had greatly enabled the city to spread well beyond the close-in suburbs. Figure 4.3 illustrates Sydney's railway system in 1894. With this railway system, concentric development of local centres began on all transport routes, and the old ribbon devel-opment was expanded. The population more than doubled between 1861 and 1881, and it more than doubled again between 1881 and 1901 with nearly half a million people in the metropolis. In 1911, the population of Sydney increased to 670,000 (Spearritt and DeMarco 1988).

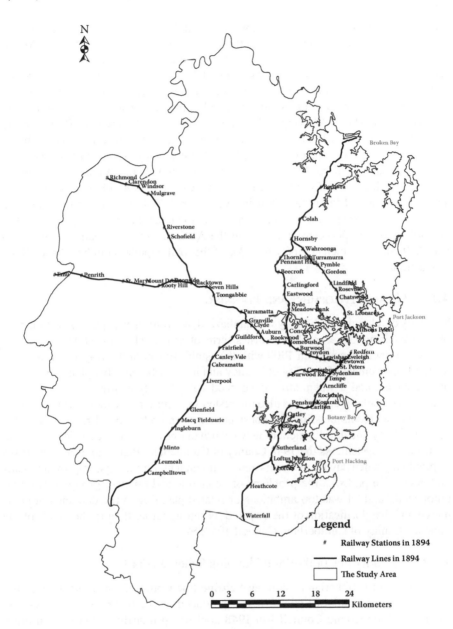

FIGURE 4.3 Sydney's railway system in 1894. (Data source: Map of the County of Cumberland New South Wales 1894. MAP RM 2824. New South Wales Department of Lands. The map is in the public domain due to age.)

Since the 1910s, the use of motor vehicles has greatly affected the development of Sydney. As the County of Cumberland Plan criticised, "uncontrolled mobility turned development patterns into a hopeless maze" (County of Cumberland Council 1948: 3). Housing estates appeared without any provision for amenities, convenient access, or related employment. Industry crept in wherever land was cheap, and little social communities became a characterless part of the great urban sprawl (Spearritt and DeMarco 1988; County of Cumberland Council 1948). By 1925, Sydney became a metropolis with a population of 1 million. By 1947, about 1.7 million people lived in the CCR, 1.2 million of whom lived in the main urban area between Port Jackson and the Georges River. Most of the remaining population lived in the Parramatta–Hornsby–Mosman triangle and in Manly–Warringah. The more remote coastal areas of Pittwater and Sutherland were sparsely settled, and the outer western and south-western suburbs were fringed by a wide belt of scattered development. Beyond the city, all the large towns such as Penrith, St. Mary's, Blacktown, Richmond, and Windsor had populations of less than 5000 (County of Cumberland Council 1948). Figure 4.4 shows the spatial expansion of the urban areas in Sydney from 1856 to 1947. In the following six decades, the population of Sydney increased to more than double the size in 1947. According to the 2006 census data from the ABS, the total population in this area was 3.69 million, which comprised 56.3% of the total population in NSW and 18.6% of Australia.

4.1.2 Urban Development and Planning

The early development of Sydney was haphazard, with only a brief period of attempted planning during Governor Macquarie's term of office. The first attempt to guide Sydney's growth occurred in 1909 with the establishment of the Royal Commission for the Improvement of the City of Sydney and its suburbs. But this plan was for city beautification, and it was mainly concerned with the city centre from the Central Station to Circular Quay. It was also interested in improving Sydney's transport systems, such as recommending the construction of an underground railway loop for the central city, the development of new suburban lines, and the electrification of the suburban rail system. As the 1948 County of Cumberland Plan criticised, "the tasks implied only a new layout of streets and of monumental buildings, improving vistas and decorative parks, ... the problems of economic and efficient grouping were all untouched," and "it was too ambitious or related planning to the demands of commerce and the glorification of the city and its setting rather than to the needs of the people" (County of Cumberland Council 1948: 8).

4.1.2.1 County of Cumberland Planning Scheme (1948)

The County of Cumberland Planning Scheme (the County Plan, in short) was the first metropolitan plan of Sydney. It was produced by the planning authority—the County of Cumberland Council—in 1948, and was primarily a plan for 2 million people. As there were already 1.7 million people in 1947, this plan was not proposed for major urban growth. Rather, it aimed at improving living conditions by applying the accepted planning principles of the time (Winston 1957).

With the assumption that secondary industry was and would continue to be Sydney's biggest employment group, the County Plan proposed an outward movement of industrial development to the suburbs, particularly to the outer suburbs in the west and south-west of the city. This was to avoid a large concentration of employment at the centre of the city and the lack of employment in the outer suburbs, and to avoid the waste of time, money, and human effort caused by long journeys to work and congestion at the city centre. The planned suburbs were to be developed as self-contained district centres.

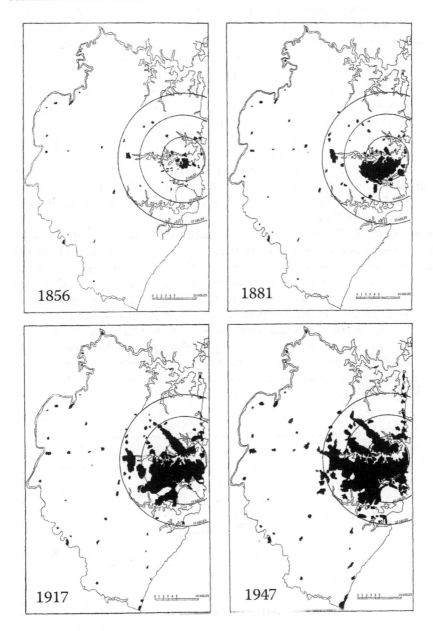

FIGURE 4.4 Sydney's urban expansion from 1856 to 1947. (From County of Cumberland Council, The Planning Scheme for the County of Cumberland, New South Wales. The Report of the Cumberland County Council to the Hon. J. J. Cahill, M.L.A. Minister for Local Government, New South Wales Department of Planning, Sydney, 1948. With permission.)

The County Plan proposed that Sydney be of a fixed size, bounded on the west by a green belt, which was to contain the city to a planned population and to prevent its outward growth (County of Cumberland Council 1948). It also identified new transport routes and defined areas that should be designed as public open spaces, district centres, and new industrial areas, taking into account as far as possible the existing land-use patterns. The County Plan was approved by Parliament in 1951 and was given statutory authority for development (Spearritt and DeMarco 1988).

After 20 years of development, the problems of the County Plan became obvious. The main setback was that the scale of growth that occurred after the County Plan was implemented was far greater than was predicted, thus causing a series of problems. The County Plan predicted that the population in the County of Cumberland would be 2,227,000 in 1972 and 2,297,000 in 1980, and some time after that there would be a tendency for the level to decline. What happened in reality was that the population in the County of Cumberland increased to 2,227,000 in 1960, which was 12 years earlier in reaching the predicted amount. By 1971, Sydney's population increased to 3 million; the actual net increase was more than twice the number predicted by the County Plan.

Consequently, the County Plan underestimated the extraordinary number of blocks of land needed to meet the new demand. Because of the average household size falling sharply from 4.0 persons per household in 1947 to 3.32 in 1971 (Figure 4.5), the population of the city of Sydney and the eastern and inner suburbs declined by about 80,000 from 1947 to 1966, although the number of dwellings increased by 14,000 over the same period. The greatest population growth had taken place in the outer areas, at Liverpool and Campbelltown to the south-west, and in an almost unbroken corridor along the western railway to beyond Penrith. With the railway lines to Lithgow and Gosford becoming electrified in 1966, some 60,000 people were living in the urban areas of the Blue Mountains and Gosford districts, many of them travelling to work in Sydney.

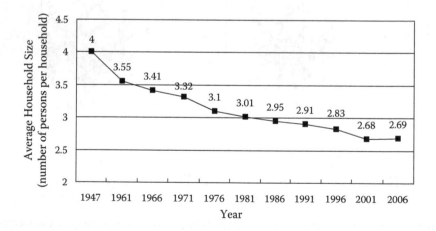

FIGURE 4.5 The average household size in Sydney from 1947 to 2006. (Data source: Census products between 1947 and 2006, Australian Bureau of Statistics.)

4.1.2.2   Sydney Region Outline Plan (1968)

In 1964, a new government authority called the State Planning Authority of New South Wales was created. This Authority began reviewing the County Plan in 1966 and started preparing for a new metropolitan plan. This became the second metropolitan plan of Sydney—the Sydney Region Outline Plan (the Outline Plan, in short). Based on the assumption that there would be a long-term trend of strong population growth, supported by the government through its immigration programme, the Outline Plan was proposed under the projection that the population in Sydney would grow from 2.7 million in 1967 to about 5.45 million in 2000. Proposals were advanced to accommodate the additional 2.75 million new residents and to plan living and working places for the new workers and the way they would travel to work (New South Wales State Planning Authority 1968).

To accommodate the new residents, development was proposed in the Outline Plan to encourage 0.5 million people to relocate to areas outside the Sydney region. It also extended the Sydney region itself to the Gosford–Wyong areas. More importantly, it provided for urban expansion along a number of growth corridors or centres within Sydney—to the north, west, south-west, and even to the north-west at a later stage. Figure 4.6 illustrates the different categories of land proposed for development based on the Outline Plan.

With the 1968 Outline Plan, development programmes along the proposed corridors and regions were implemented in the following years. For instance, by 1979, major public initiatives took place at Macarthur in the south-west sector as the growth centre; the population doubled, and it became an area with the highest growth rate in NSW. Industrial and commercial sites were developed, and there were notable successes in the growth of employment and services in Campbelltown and Mount Druitt. In the west sector, considerable population growth occurred, although growth in employment and improvements in community facilities and communication networks were limited (New South Wales Planning and Environment Commission 1980).

Although development did occur in the proposed corridors and centres, it was not at the predicted rate. This was largely due to the overestimation of population growth. In practice, the population growth declined significantly in the 1970s due to lower birth rates and reduced immigration. Therefore, the total population was well below the number projected in the Outline Plan. As the 1980 review of the Department of Environment and Planning pointed out, "because of the priority attached to accommodating a rapid expansion of residential development, it emphasized growth and urban extension to a greater extent than management of existing urban areas" (New South Wales Planning and Environment Commission 1980: 22). In the existing urban areas, the population declined over the decade by 60,000, even though a significant amount of redevelopment occurred. The decline was mainly in the inner areas of the city.

Another weakness of the Outline Plan was the analysis of transport, which had been done without the essential data needed for land-use and transport recommendations (New South Wales Planning and Environment Commission 1980). Nevertheless, the Outline Plan was an important document in planning Sydney's urban development because it nominated corridors and areas where growth could take place (Spearritt and DeMarco 1988).

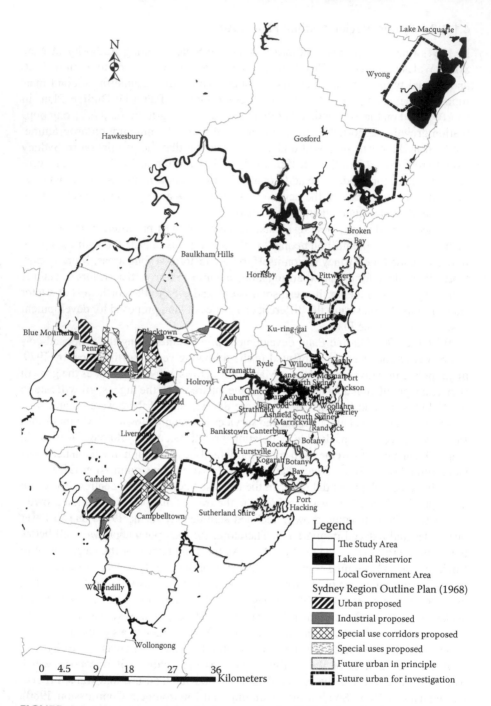

FIGURE 4.6 Areas proposed for development in the 1968 Sydney Region Outline Plan. (Data source: New South Wales State Planning Authority 1968.)

4.1.2.3 Sydney into its Third Century (1988)

By 1986, the Sydney region had a population of approximately 3.47 million. As a result of the shortage of land and spiralling land costs in the 1980s, a new metropolitan strategic plan called *Sydney into its Third Century* was released (New South Wales Department of Environment and Planning 1988). The target of this plan was to provide Sydney with a framework for future development and redevelopment within which local decisions could take place. With the understanding that Sydney would retain its role as a major financial business centre in Australia and as a destination for overseas migrants, it was expected that this region would keep growing, although the rate of growth might change. An estimation was made that, by 2011, the population in the Sydney region would be around 4.5 million, although this level of population might be reached sooner or later depending on net migration. The metropolitan plan was designed to accommodate 4.5 million people in Sydney irrespective of when the number was reached (New South Wales Department of Environment and Planning 1988).

Two alternative strategies for the development of Sydney were formulated. One was the concentrated strategy, referring to the concentration of population and employment, and the other was the dispersed strategy, referring to the continuation of the trends in population and locations of employment of the 1980s. By comparing the advantages and disadvantages of the two alternatives, the concentrated strategy was considered more cost effective for the government and the community, decreasing potential air and water pollution and conserving land for future urban, agricultural, and recreational purposes. With the concentrated strategy, housing densities in new areas would be increased from 8 to 10 residential lots per hectare. Such a target meant an increase of about 25% in population per hectare in new areas, and some 13,000 hectares would be saved from urban development to house an extra 1 million people. As the foundation stone of this strategy, urban consolidation in the established areas was to be implemented by both the public and private sectors by one of the following means:

- Redevelopment in some low-density areas
- In-fill on vacant or non-residential sites
- Conversion of non-residential buildings to residential houses
- Retention of existing housing under threat of either demolition or conversion to non-residential uses (New South Wales Department of Environment and Planning 1988)

Apart from urban consolidation in the established areas, the 1988 Metropolitan Plan also proposed new growth sectors for future urban expansion. These sectors were identified by considering the physical, social, and economic constraints on land development. Figure 4.7 shows the areas proposed for future urban development in the 1988 and 1995 urban planning schemes. The proposed urban areas that are outside the study area of this book are not illustrated in the figure.

The 1988 Metropolitan Plan was an important document that guided Sydney's development. However, in practice, urban consolidation in established areas was limited.

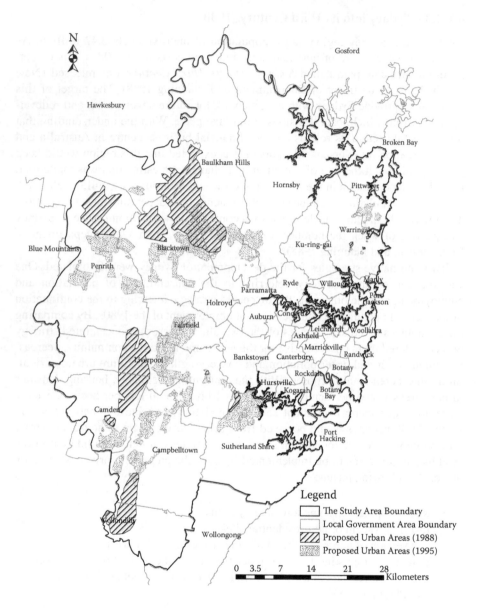

FIGURE 4.7 Areas proposed for future urban growth in the 1988 and 1995 urban planning programmes. (Data sources: New South Wales Department of Environment and Planning 1988; New South Wales Department of Planning 1995.)

Only about 30% of all new dwellings constructed in the established areas were multi-unit dwellings. The growth of Sydney took place beyond the reach of existing rail systems, which exacerbated reliance on private motor vehicles.

4.1.2.4 Cities for the 21st Century (1995)

As Sydney approached the 21st century, more attention was given to sustainable development and environmental protection. From 1992 to 1995, the New South Wales Government Department of Planning began to propose a new strategy for planning Sydney's future development. Based on a series of publications proposed by the department, including *Updating the Metropolitan Strategy* (New South Wales Department of Planning 1992), *Sydney's Future* (New South Wales Department of Planning 1993a), and *Integrated Transport Strategy for Greater Sydney* (New South Wales Department of Planning 1993b), a new strategy for planning the future of the Greater Metropolitan Region of Sydney, Newcastle, the Central Coast, and Wollongong—*Cities for the 21st Century*—was published (New South Wales Department of Planning 1995). Based on new demographic projections incorporating lower migration assumptions, the population of Sydney was expected to be 4.48 million by 2021, about 10 years later than was anticipated in the 1988 Metropolitan Plan. This would require the provision of approximately 520,000 new dwellings by 2021.

Built upon the 1988 Metropolitan Plan of Sydney, the new strategic plan puts an even stronger emphasis on containing the expansion of major urban centres. It was planned that approximately two-thirds of Sydney's new housing would be built in the established areas, and only one-third in "greenfield" estates. By increasing the proportion of new dwellings built each year in multi-unit form and increasing the average neighbourhood dwelling density of new greenfield housing estates, the newly built-up areas were expected to be more compact than the established areas. According to this plan, around 30% of new housing would be located in the west and north-west sectors; 43% in the south-west sector, including Fairfield, Liverpool, Campbelltown, and Camden combined; 7% in Wyong on the north-east, which is outside the County of Cumberland boundary; and the remaining 20% in a range of other fringe areas. For the longer term, when an expansion of Rouse Hill would be needed, areas close to the Riverstone rail line would be investigated. An outline of a number of areas proposed in the 1995 Metropolitan Plan for future urban development is also illustrated in Figure 4.7.

4.1.2.5 City of Cities (2005)

In December 2005, the New South Wales government Department of Planning released a new Metropolitan Strategic Plan, the *City of Cities*. Based on a growth anticipated scenario that Sydney's population will increase by 1.1 million people from 4.2 million in 2004 to 5.3 million in 2031, this strategic plan provides a broad framework to facilitate and manage growth and development in Metropolitan Sydney over the next 25 years (New South Wales Department of Planning 2005). Guided by the principles of economic, social, and environmental sustainability, the Metropolitan Strategic Plan aims to achieve a more sustainable city by enhancing liveability, strengthening economic competitiveness, ensuring fairness and access to jobs and services, protecting the environment, and improving governance.

There are eight key elements identified in the plan that illustrates the NSW government's vision for Sydney in 2031. These key elements include the following:

1. Stronger cities within the metropolitan area
2. Strong global economic corridor
3. More jobs in western Sydney
4. Contain Sydney's urban footprint
5. Major centres emerge as job, service, and residential locations
6. Fair access to housing, jobs, service, and open space
7. Connected centres
8. Better connected and stronger regions (New South Wales Department of Planning 2005)

A distinct feature of this strategic plan is its centre-focused strategy on urban development. In this plan, Sydney was defined as one global city with five key regional cities—Parramatta, Liverpool, Penrith, Sydney CBD, and North Sydney. In addition, a hierarchy of centres, including major and smaller centres, was also identified. Figure 4.8 illustrates the hierarchy of cities and growth centres as well as major corridors for development proposed in the 2005 Metropolitan Strategic Plan.

According to this plan, an *existing major shopping and business centre* is one with council offices, taller office and residential buildings, a large shopping mall and central community facilities for the district; some *potential major centres* are also planned. Such major centres will have improved transport links with fast, safe, and reliable train services and a network of strategic bus corridors connecting the centres across the city.

Some *smaller centres* with good public transport will be managed by local government. These include town centres, villages, and neighbourhood centres. A *town centre* is a large group of shops and services with one or two supermarkets, sometimes a small shopping mall, and some community facilities such as a local library, a medical centre, and a variety of specialist shops. The extent of a town centre is approximately an 800 m radius, which is widely accepted as a comfortable 10-min walk. A *village* is a strip of shops for daily shopping, which typically includes a small supermarket, butcher, hairdresser, restaurants, and take-away food outlets. The extent of a village centre is approximately a 400–600 m radius. A *neighbourhood centre* is a small group of shops that people can walk to and buy commodities such as milk and newspapers. Any street with a corner shop can be considered as a neighbourhood centre. The extent of a neighbourhood centre is approximately a 200 m radius.

Over three-quarters of new housing is planned to be located in the strategic centres, smaller centres, and corridors, which can significantly increase the population and housing density around them. In addition, the Metropolitan Strategic Plan also identifies two land release areas, one in the north-west and one in the south-west, as is shown in Figure 4.8. These areas are planned to be developed as new growth centres in the near future.

The impact of the 2005 Metropolitan Strategic Plan on Sydney's urban development will be seen in the coming years. By applying the fuzzy constrained cellular automata model, which is to be configured and calibrated in the next chapter of the book, a simulated scenario of future urban development under this metropolitan strategic plan will be presented in Chapter 6.

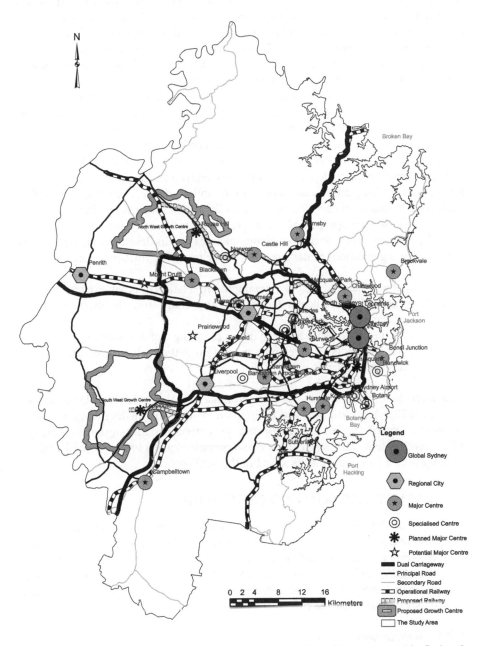

FIGURE 4.8 The hierarchy of cities, growth centres, and corridors proposed in Sydney's Metropolitan Strategic Plan (2005). (Data source: New South Wales Department of Planning 2005.)

4.1.3 Issues Relating to Sydney's Urban Development

Although the early development of Sydney was haphazard, attempts have been made in urban planning since the early 20th century, especially after World War II. When looking back at the urban development and planning of Sydney since 1948, significant gaps exist between various planning schemes and practical urban development. Although the County of Cumberland Plan contributed to the decentralisation of industries and the development of new district centres, it failed to estimate the correct population growth. As a consequence, the plan underestimated the extraordinary number of blocks of land necessary to meet the development demand. It also constrained the development of urban areas within a fixed size, bounded on the west by a green belt, although, in practice, development still took place outside such bounded areas, resulting in the rezoning of a large proportion of the green belt as residential land.

In contrast to the first urban planning scheme, the second metropolitan plan of Sydney—the Sydney Region Outline Plan—overestimated population growth. The main target of this plan was the population doubling in size, even though this plan nominated a number of corridors and centres for future urban expansion.

When reviewing the process of urban development of Sydney since 1976 and evaluating the effectiveness of various planning schemes on this development, the answers to the following questions remain unclear, hence requiring further research:

1. In what way has Sydney been developing in relation to urban expansion?
2. What are the fundamental factors or rules controlling the expansion of urban areas in Sydney?
3. What is the relationship between urban planning and development?
4. How has Sydney been developing in relation to the various urban planning schemes since the mid-1970s?

To answer these questions, the cellular automata model of urban development constructed in Chapter 3 will be tested to simulate Sydney's urban development. Before the model can be applied and calibrated, geographical data relevant to Sydney's urban development need to be collected and processed using various spatial analysis techniques in a GIS, which is discussed in the next section.

4.2 DATA COLLECTION AND PROCESSING

Geographical data were required to generate an information system to illustrate the spatio-temporal changes of the urban areas of Sydney. The data were available from different agencies, including Geoscience Australia, the Australian Bureau of Statistics (ABS), and some other government departments. Various spatial analysis techniques in a GIS were applied to process the raw data and verify the quality and consistency of data collected from different sources.

4.2.1 Topographic Data

GEODATA TOPO-250K (Geoscience Australia 2006) produced by the Australian Surveying and Land Information Group (AUSLIG), the predecessor of the National Mapping Division of Geoscience Australia, was used in this research.

This product contains a medium-scale vector representation of the topographic features of Australia, including contours and spot elevation; drainage networks including watercourses, lakes, wetlands, and offshore features; and infrastructure data including roads, railways, and associated structures, along with localities and urban built-up areas. The data were originally digitised from the 1:250,000 National Topographic Map in 1994. However, it was subsequently revised in its Series 2 and Series 3 products by incorporating features using satellite imagery and other supplementary information as revision sources. It is claimed that data contained in the Series 3 products have a currency of less than 5 years in any location (Geoscience Australia 2006).

Relief data of Sydney were extracted from this data package. The point data of spot elevations were interpolated in ArcGIS™ using a linear interpolation technique, also incorporating data on water bodies and the coastline to generate a digital elevation model (DEM) and a slope data layer. Both the DEM and slope data were used in the cellular automata model as constraints to Sydney's development.

4.2.2 TRANSPORTATION NETWORK

Data illustrating the transportation network up to the year 1994 were included in the TOPO-250K Series 1 data set. In the Series 3 data release, a number of new transportation lines constructed in the 1990s, including the Hume Highway and the Western Freeway, were added to the database, making it possible to build up a temporal database of the transportation network of Sydney from 1994 to 2006.

The major transportation network in Sydney includes both roads and railways. Based on the railway network constructed over a century ago, data illustrating the latest update on the railway system were gathered from the TOPO-250K Series 3 data set. Although the majority of railway lines run on multiple tracks, a small portion are running on single tracks. All the latest railway lines and stations were identified in the data set as well.

Roads in the GEODATA TOPO-250K Series 3 product have been classified into five categories based on their widths and types. These include Dual Carriageway (DC), Principal Road (PR), Secondary Road (SR), Minor Road (MR), and Vehicle Track (VT). For the purpose of modelling Sydney's urban growth, only the DCs, PRs, SRs, and some MRs are used as input data layers. Figure 4.8 also includes the current and proposed transportation network in Metropolitan Sydney.

Although there is no data available to illustrate the temporal change of transport network from 1976 to 1994, research on how the transportation network of Sydney had been changing over that time period was undertaken. The major frame of the transportation network, including railways, arteries, and carriageways, as well as major ferry routes in Sydney, was set up in the 1950s and 1960s or even earlier (New South Wales Department of Planning 1993b). Hence, data illustrating the transportation network in the early 1990s were used as the initial road data to feed the model when transition rules representing the impact of transportation network on Sydney's urban development was implemented. Subsequently, new roads and railways constructed during the 1990s were introduced into the model based on their completion dates. In addition, roads completed after 2006 as well as railway lines currently under construction or proposed for future construction will be used when the model is applied to simulate future urban scenarios under the development planning conditions.

4.2.3 PHYSICAL URBAN AREAS

Similar to the topographic and transportation network data, data illustrating the urban built-up areas of Sydney were also available from the TOPO-250K data package. However, as the data contained in this data set have a currency of less than 5 years (Geoscience Australia 2006), this data source cannot be used to illustrate the temporal change of urban areas over the past three decades. Therefore, another source of data from the ABS was used instead.

The ABS is the official government organisation in Australia that is responsible for collecting and publishing census data of population and housing at every 5-year interval since 1961. Hence, the census data can be used to set up temporally consistent standards defining urban areas. As discussed in Chapter 3, instead of defining a sharp boundary between urban and non-urban areas, the census data can be used to define a fuzzy concept between the two, which would help to model the fuzzy process of non-urban to urban conversion.

The basic spatial unit for collecting census data is the Census Collector's District (CD or CCD), which is the smallest geographical area defined in the Australian Standard Geographical Classification (ASGC). This spatial unit also serves as the basic building block for the ASGC, and is used for the aggregation of statistics for larger ASGC areas (Australian Bureau of Statistics 2006). CDs are generally designed to form the workload for a census collector in the field (7 days before and 12 days after the census night). On an average, there are about 250 dwellings in each CD in the urban areas, whereas in rural areas the number of dwellings per CD reduces as population densities decrease. The sizes of CDs vary significantly from one CD to another as a result of the changes in population density.

Although census data have been published at CD level since 1966, no data illustrating the geographical boundaries of CDs in digital form for the earlier census data from 1966 to 1976 were found; therefore, it was difficult to compute either population or dwelling density at the CD level using the census data. However, by using a comparability indicator in the 1981 Census indicating whether changes had been made to the CD boundaries over the two census periods between 1976 and 1981, it was possible to trace back the 1976 CD boundaries from the 1981 CD boundaries. For those incomparable CDs, the printed CD boundary maps of 1976 were consulted thereafter. Consequently, the starting date for the modelling of the urban development of Sydney is set to 1976, which is the earliest date with reliable data to set up the initial state of cells of the cellular automata model.

For the latest census data from 1981 to 2006, both the CD boundaries and statistics data were released by ABS in a digital format, which was processed in GIS to define the urban extents of Sydney from 1981 to 2006. Section 4.3 presents the methodology used in defining the urban extent based on the fuzzy set concept.

4.2.4 LAND EXCLUDED FROM URBAN DEVELOPMENT

Another data layer used in the model illustrates areas excluded from urban development. Areas such as water bodies (including lakes, reservoirs, and water courses), water supply reserves, forestry reserves, nature conservation reserves, prohibited areas for defence purposes, mines, or cemetery areas where urban development is not permitted are termed *excluded areas*, that is, they are crisp barriers to urban development. This data layer was initially collected from various sources, including the AUSLIG, the New South Wales Department of National Parks and Wildlife Service, the Sydney Water Corporation, and the State Forest of New South Wales. With the release of the GEODATA TOPO-250K Series 3 product, the data were consolidated into a few data layers that were processed in GIS as one input data layer of the model.

There are two types of areas excluded from urban development: areas that are water bodies, national parks, forest reserves, nature conservation reserves, and water supply reserves. These areas are considered as hard constraints to urban development, that is, it is not possible for them to be developed into urban areas. Figure 4.2 contains a data layer illustrating areas excluded from urban development in Metropolitan Sydney and its surrounding regions.

The other type of restricted land, including cemeteries within urban built-up areas, golf courses, aerodromes, military camps, and lands reserved by the commonwealth government for various purposes, is also a strong barrier to urban development under current conditions. However, these areas could be developed at some stage under certain circumstances, such as when population increases significantly and there is shortage of land for urban development. Therefore, this type of land may not be a crisp barrier to urban development, and so, it is termed *semi-excluded land*. Once the semi-excluded land is released for urban development, development will be dealt with in the same way as in other areas available for development. If only a part of a semi-excluded area is released for development, the constraint on the rest of the land still exists until it is fully released.

4.2.5 URBAN PLANNING SCHEMES

As previously discussed, a number of urban planning schemes have been implemented in Metropolitan Sydney since the 1940s. The planning schemes may have influenced the urban development of Sydney to some extent, or they may not have influenced its urban growth at all. In order to evaluate whether or to what extent urban planning can affect urban development, some secondary transition rules were proposed in the cellular automata model to represent the impact of this institutional factor. Apart from an understanding of the overall urban planning policies, data illustrating areas proposed for future urban development were also collected from planning schemes, such as the Sydney Region Outline Plan (1968), Sydney into its Third Century (1988), the Cities for the 21st Century (1995), and the City of Cities (2005). Figures 4.6–4.8 show the various areas planned for urban development under these planning schemes. The urban planning schemes proposed in 1968, 1988, and 1995 were considered to have covered an effective period for the cellular automata model to operate from 1976 to 2006. Hence, they were used to calibrate the model and evaluate how the urban areas of Sydney had been developing in relation to urban planning policies and programmes. The latest Metropolitan Strategic Plan published in 2005 is to be applied to generate future urban scenarios from 2006 to 2031.

4.3 DEFINING SYDNEY'S URBAN AREAS WITH FUZZY SET THEORY

4.3.1 URBAN AREA CRITERIA FOR STATISTICAL PURPOSES

In the 1960s, Linge (1965) developed a set of principles and criteria to delimit urban centres in Australia at the request of the ABS for statistical purposes. According to Linge (1965), urban area boundaries can be defined at the CD level through pre-census preparations and post-census analysis. At the pre-census preparation stage, an initial distinction between urban and rural CDs was obtained by calculating the density of population per square mile. Those CDs with a population density of at least 500 persons per square mile were classified as urban areas, and those with a density lower than 500 persons per square mile were classified as rural (Linge 1965: 67). In addition to the density criterion, the following situations were considered to split a CD into two or more CDs at the pre-census stage:

- "The splitting CDs on the periphery of the urban area, regardless of the density, if (i) there is a minimum population of 500, and (ii) there is a gap of more than a mile between the edge of urban development and the further edge of the CD" (Linge 1965: 71–72).
- "Where there is a gap of less than two miles (shortest road distance) from the edge of one area of urban development to another area of urban development the gap should be ignored and the urban areas should be regarded as continuous" (Linge 1965: 73).
- "Where a CD consists of an industrial area (which would be classified as urban) and a rural area (which would not be counted as urban)" (Linge 1965: 75).

A number of rules were proposed to delimit urban areas during the post-census analysis stage. These rules include

- "A CD with a density of over 500 persons per square mile which is contiguous with an urban area will be included as part of that urban area" (Linge 1965: 80).
- "A CD with a density of less than 500 persons per square mile which is completely surrounded by CDs with over 500 persons per square mile will be included as part of the urban area" (Linge 1965: 81).
- "Any area previously classified as urban should be again classified as urban even if the CD concerned has a density of less than 500 persons per square mile" (Linge 1965: 81).
- Some low density CDs (less than 500 persons per square mile) but containing special land uses and being contiguous to the periphery of the urban areas may be included in an urban area depending on the type of land use and the nature of the areas surrounding the CDs (Linge 1965).
- Outlying urban areas were defined as areas which "should have a population of at least 1,000" (Linge 1965: 84).

- Holiday settlements were classified as urban areas where "(i) there are at least 250 dwellings altogether and at least 100 dwellings were occupied on census night, and (ii) the settlement has a recognisable 'core'" (Linge 1965: 90).

The basic criteria developed by Linge (1965) were accepted by ABS with subsequent amendments at the Statistician's conferences in 1965 and 1969 and at the Review of ABS Statistical Geography in 1988 (Australian Bureau of Statistics 2005). The core points of the delimitation criteria for urban centres currently in force in the ASGC are as follows: For urban centres having a population of 20,000 or more, an urban boundary is defined that consists of a cluster of contiguous CDs and other urban areas. CDs classified as urban include the following:

- All contiguous CDs that have a population density of 200 or more persons per square kilometre shall be classified as urban.
- A CD consisting mainly of land used for factories, airports, small sports areas, cemeteries, hostels, institutions, prisons, military camps, or certain research stations shall be classified as urban if contiguous with CDs that are themselves urban.
- A CD consisting mainly of land used for large sporting areas, large parks, explosives handling and munitions areas, or holding yards associated with meatworks and abattoirs shall be classified as urban only if it is bordered on three sides by CDs that are themselves classified as urban.
- Any area completely surrounded by CDs that are urban must itself be classified as urban.

For urban centres with a population between 1,000 and 19,999, boundaries are delimited subjectively by the inspection of aerial photographs, by field inspection, and/or by consideration of any other information that is available. All contiguous urban growth is to be included (even if this would not necessarily occur if the density criterion was applied), together with any close but non-continuous development that could be clearly regarded as part of the urban centre. However, for urban centres containing a population approaching 20,000, the objective criteria applied for urban centres with 20,000 people should also be considered (Australian Bureau of Statistics 2005).

4.3.2 DEFINING A FUZZY BOUNDARY OF SYDNEY'S URBAN AREAS

As Linge's criteria were not intended to "draw a boundary around the built-up area of a town, but around the areas in which people are living an urban way of life" (Linge 1965: 67), the 500 persons per square mile or approximately 200 persons per square kilometre criterion is very low for delimiting an urban area when compared to similar criteria applied in other countries (see Chapter 3, Section 3.1.3.1). Even in Sydney, many areas that meet the density criterion of 200 persons per square kilometre are still in a rural state to a large extent. Examples can be found around Hornsby and Ku-Ring-Gai areas in the northern part of Metropolitan Sydney. Through the application of a fuzzy membership function, criteria based on fuzzy sets can be defined to illustrate the extent of the urban areas of Sydney.

In Chapter 3, Section 3.1.3.2, a fuzzy membership function was defined in Equation 3.10 to delimit the urban area in a fuzzy set. This equation can be rewritten as Equation 4.1.

$$\mu_{S_{urban}}(x_{ij}) = \begin{cases} 0 & \rho_{x_{ij}} < \rho_0 \\ \dfrac{\rho_{x_{ij}} - \rho_0}{\rho_1 - \rho_0} & \rho_0 \leq \rho_{x_{ij}} < \rho_1 \quad (x_{ij} \in X) \\ 1 & \rho_{x_{ij}} \geq \rho_1 \end{cases}$$

(4.1)

where $X = \{x\}$ denotes an urban fuzzy set; $\mu_{S_{urban}}$ is the membership function of cell x_{ij} in the urban fuzzy set; and ρ_0 and ρ_1, respectively, are the lower and upper thresholds to delimit the urban extent.

From Linge's criteria of urban areas and the criteria currently in force in the ASGC, it was accepted that a CD with a population density of less than 200 persons per square kilometre could be regarded as non-urban unless it met one of the other subcriteria as outlined in Section 4.3.1. Therefore, the value 200 persons per square kilometre can be used as the lower threshold in defining the urban fuzzy set. If an area has a population density of less than the lower threshold of 200 persons per square kilometre, it has no membership in the urban fuzzy set, or the membership value of this area in the urban fuzzy set is 0.

However, defining the upper threshold of the membership function in the urban fuzzy set is not simple. Various criteria can be applied to measure whether an area can be considered as a fully developed urban extent. These criteria may include population density, dwelling density, land-use type, availability of infrastructure supply, as well as some other socio-economic indicators. Pragmatically, it was borne in mind that "a fully built-up housing area in a large Australian town would have a density of at least 3,000 persons per square mile" (approximately 1200 persons per square kilometre; Linge 1965: 67).

With the constant growth of urban population and the limited supply of land, the lot sizes for residential dwellings are getting smaller on new urban residential subdivisions, and some existing urban areas have also undergone an urban consolidation process. This resulted in high population densities even though the average household size is getting smaller. For instance, in the concentrated strategy of the 1988 Metropolitan Plan, it was proposed that the housing densities in new areas should be increased from 8 to 10 residential lots per hectare (New South Wales Department of Environment and Planning 1988). Since the 1990s, local government authorities started to allow additional dwellings of a suitable type to be developed within the existing urban areas in order to minimise urban sprawl. The most recent Sydney Metropolitan Strategy also planned for about 30–40% of new housing to be constructed in new release land, and the remaining 60–70% in existing urban areas (New South Wales Department of Planning 2005). Consequently, the population density criteria used in measuring the extent of urban development needs to be adjusted over time.

With this development and change in place, it is suitable to propose two different population density thresholds to quantify as the minimum value to define the fully urban built-up area of Sydney; they are 1200 persons and 1500 persons per square kilometre. The lower value of 1200 persons per square kilometre was applied to define urban areas in the 1970s and 1980s, and the higher value of 1500 persons per square kilometre was applied to define urban areas from the 1990s onward. Areas with a population density between 1200 and 1500 persons per square kilometre, which have already been defined as fully urban areas in an earlier census, remain as fully urban during the model simulation process; such lower-density urban areas may be subject to urban redevelopment or consolidation should there be any driving force for urban consolidation at later stages. However, the process of urban redevelopment or consolidation is not considered in the current version of the model. For new areas developed since the 1990s, the higher density value of 1500 persons per square kilometre applies when they are considered as fully urban areas. Therefore, parameter ρ_1 in Equation 4.1 is fed with these two threshold values to define the urban states for different time periods.

Areas having a population density between 200 and 1200 or 1500 persons per square kilometre, or a dwelling density equivalent to the population density, are defined *as partly urban areas,* with their membership grades in the urban fuzzy set being determined by Equation 4.1. Using this criterion, the urban areas of Sydney in 1976, 1981, 1986, 1991, 1996, 2001, and 2006 were produced and stored in digital format in a GIS, which are used to calibrate the cellular automata model and simulate its urban development over time.

4.3.3 Visualising Sydney's Urban Development in Space and Time

GIS has been applied widely to visualise the spatio-temporal process of physical changes (Bell, Dean, and Blake 2000; Batty 1998, 1994; Clarke and Gaydos 1998; Batty and Howes 1996). This technology has also been used in this book to visually explore the development of urban areas of Sydney over the period 1976–2006. By overlaying data layers illustrating the urban areas generated from the censuses on the topographic data layer, a temporal process of urban expansion of Sydney in relation to relief was visualised, and the spatial change of urban areas over different time periods was detected. Urban expansion was visualised in relation to the transportation network or other infrastructure, or in relation to areas planned for development in the urban planning schemes. The influence of infrastructure or urban planning on urban growth, or vice versa, could therefore be evaluated.

The analysis and visualisation of the census data in GIS demonstrate that the fully urban areas of Sydney had increased from 16.5% in 1976 to 23.7% in 2006. Whereas some non-urban areas had developed into partly urban areas, some partly urban areas had also developed into fully urban areas. Therefore, the proportion of partly urban areas in the total area of the region increased only slightly from 12.0% in 1976 to 15.1% in 2006 (Table 4.1 and Figure 4.9). Overall, an area of 533.5 km² had been developed from non-urban to partly urban or urban, or from partly urban to urban. This resulted in an average urban expansion rate of 1.03%.

TABLE 4.1

Change of Urban Area in Sydney from 1976 to 2006

2006 ╲ 1976	Non-urban	Partly urban	Urban	Total
Non-urban	2411.4|0	0	0	2411.4
Partly urban	252.1	340.2	0	592.3
Urban	149.9	131.5	650.3	931.7
Total	2813.4	471.7	650.3	3935.4

Note: Unit: square kilometre.

Compared with its average annual population growth rate of 0.8% over the same period, the rate of urban expansion in Sydney was greater than that of its population growth.

Visualising the urban change in Sydney spatially (Figure 4.10), it appears that the urban expansion of Sydney from 1976 to 2006 was largely in accordance with what was planned in the Sydney Region Outline Plan proposed in 1968 (New South Wales State Planning Authority 1968; see Figure 4.6). The urban development mainly occurred at the south-west sector covering Campbelltown, Camden, and Appin; the west sector covering Penrith, Blacktown, and Fairfield-Hoxton Park; and the south sector at Liverpool and Sutherland. Development also occurred in areas to the north-west along the roads and railway lines, and in the north-east in Warringah–Pittwater. Whether the Sydney Region Outline Plan and subsequent planning schemes had successfully controlled or influenced the urban development of Sydney over that time period, or whether they had just identified the obvious directions of urban growth in the area, is worth further consideration.

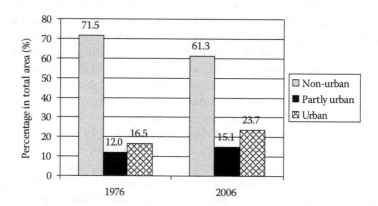

FIGURE 4.9 Urban area changes in Sydney from 1976 to 2006. (Data source: Census products 1976 and 2006, Australian Bureau of Statistics.)

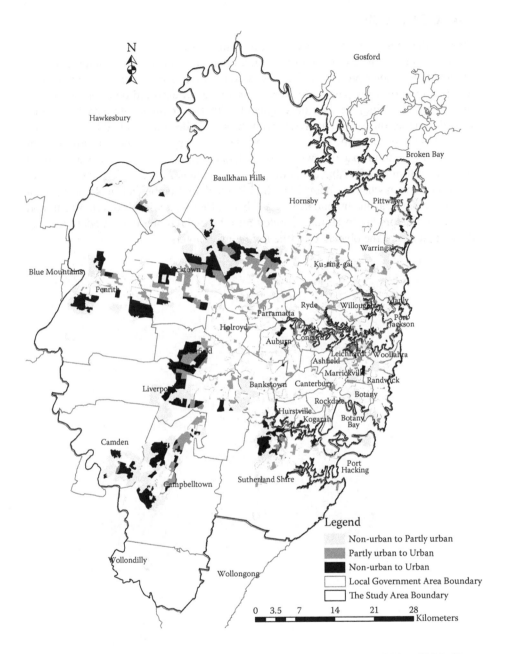

FIGURE 4.10 The spatial expansion of urban areas in Sydney from 1976 to 2006. (Data source: Census products 1976 and 2006, Australian Bureau of Statistics.)

4.4 CONCLUSION

One of the most important tasks in developing a successful modelling system is to understand the logical operation of the system and prepare data suitable for the model. This chapter presented discussions concerning Sydney's urban development and planning over the past three decades or longer. It explored data relating to the modelling of Sydney's urban development with the cellular automata technique. In addition, various spatial analysis techniques of a GIS were applied to process the raw data and visualise the urban development of Sydney over time. Visualisation of the urban growth in space shows that the urban development of Sydney was focused in areas of limited relief and along the transportation network, which was largely in accordance with what was planned in the 1968 Sydney Region Outline Plan. To what extent the physical and human factors had been affecting Sydney's development, and how these factors are to be incorporated into the cellular automata model of urban development, are the objectives of the following chapters.

5 Modelling the Urban Development of Sydney

Model Specification, Calibration, and Implementation

This chapter applies the fuzzy constrained cellular automata model of urban development constructed in Chapter 3 to the study area in Sydney to simulate its urban development from 1976 to 2006. The model was configured and calibrated using the Sydney data sets processed in Chapter 4. Section 5.1 specifies the initial settings of the model in terms of the cell size and state, neighbourhood scale, transition rules, as well as the temporal dimension of simulation intervals. Section 5.2 sets up the principles and approach for calibrating the model and assessing the accuracy of the model's outcomes. In Section 5.3, implementation of the model within a geographical information system (GIS) environment is presented, followed by conclusions on model calibration and implementation addressed in this chapter.

5.1 MODEL SPECIFICATION

5.1.1 CELL SIZE AND STATE

A regular spatial tessellation of the urban space within the Sydney Metropolitan Area as defined in Chapter 4 was first computed with a spatial scale of 250 m. That is, each cell on the model represents an area of 250×250 m^2 on the ground.

Using the fuzzy set approach, a continuous state of cells in the urban fuzzy set was defined; the state of cells was represented by the value of their membership in the urban fuzzy set. This membership value ranges from 0 to 1 (inclusive). A full membership of 1 indicates that the cell is fully developed into an urban state, whereas a 0 membership value indicates that the cell is not developed at all, and therefore it is in a non-urban state. Cells with a membership value between 0 and 1 (exclusive) have partial membership in the urban fuzzy set; therefore, they are in a partly urban state. The actual position of the cell in the partly urban state is reflected by the membership value of the cell. For example, if a cell has a membership value of 0.8, it is in a higher partly urban state than a cell with a membership value of 0.3 in the same urban fuzzy set. Details on the membership function and its application in defining the state of a cell used in the model are shown in Chapter 4, Section 4.3.2.

Other data sets presented in Chapter 4, Sections 4.2.1, 4.2.2, 4.2.4, and 4.2.5, including topographic data, transportation data, data on land excluded from urban development, as well as urban planning schemes, were also processed as regular spatial grids at a spatial scale of 250 m. These data sets are used to feed the model to simulate the urban growth in Metropolitan Sydney.

5.1.2 NEIGHBOURHOOD CONFIGURATION

With the concern that distortions may exist in a rectangular neighbourhood-type configuration (Li and Yeh 2000), a circular neighbourhood was applied; the size of the circle was set to a two-cell radius, same as the radius of the circle measured perpendicular along the x- and y-axes of the central cell. A cell whose centre was within the circular area was taken as a neighbour of the cell in question (Figure 5.1 b). Other neighbourhood sizes, namely, a small neighbourhood (with a radius of one cell), a large neighbourhood (with a radius of three cells), and a very large neighbourhood (with a radius of four cells) were also tested (Figure 5.1a, c, and d), and their impacts on the model's simulation results are discussed in Chapter 6.

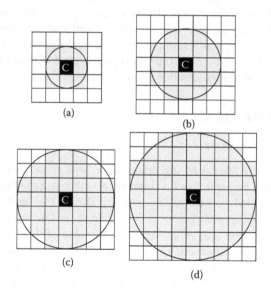

FIGURE 5.1 Different neighbourhood sizes experimented in the fuzzy constrained cellular automata model. Four sizes of neighbourhood, the radius of which ranges from one to four cells, representing small, medium, large, and very large neighbourhoods, respectively, were tested. A cell whose centre is within the circular area is taken as a neighbour of the central cell C: (a) A small neighbourhood, (b) a medium-size neighbourhood (this is the initial setting of the model), (c) a large neighbourhood, and (d) a very large neighbourhood. The cell C in black is the processing cell; the grey cells are the neighbouring cells, whereas the white cells are not.

5.1.3 TRANSITION RULES

Urban development is the result of both internal and external forces. Internally, an area tends to continue its development if it has started to develop from a rural to an urban state, especially if this natural tendency is supported by development from within its neighbourhood. Externally, factors such as the geographical conditions of the area, its socio-economic conditions, as well as institutional controls may also have an impact on the process of its development. Physical constraints such as water bodies and steep terrain may restrict or slow down the process of urban development, as do institutional controls that may accelerate or prohibit further urban growth. Socio-economic factors such as land availability and demands on available land, accessibility to nodes of employment and other services, and facilities such as schools, shops, public transport, and contiguity to existing urban areas also play important roles in urban development. These have been studied extensively in urban models (Bell, Dean, and Blake 2000; Li and Yeh 2000; Wu and Webster 1998c; Allen and Sanglier 1978; Malm, Olsson, and Wärneryd 1966). Using the cellular automata model of urban development constructed in Chapter 3, the internal and external forces affecting the process of urban development in Sydney are simulated through the setting of primary and secondary transition rules, respectively.

5.1.3.1 Urban Natural Growth Controlled by Primary Transition Rules

The primary transition rules presented in Chapter 3, Section 3.3.3.1, deal with an ideal situation in which development of an urban area (a cell in the model) has resulted from its own propensity for development (natural growth) and the support it may have received from its neighbourhood. The process of this development follows a logistic curve as described in Chapter 3, Equation 3.17. Other environmental or socio-economic factors such as topographic features, urban infrastructure, or transportation network are not considered in the primary transition rules, even though these factors may constrain or accelerate the process of an area's development. Table 5.1 presents the initial specification of parameters configured in the logistic curve and the rule-firing thresholds as defined in Chapter 3, Section 3.3.3.2.

TABLE 5.1
The Initial Specification of Parameters in the Model

Model's initial settings	Value of parameters
The time frame for a medium-speed urban development process ranges from 15 to 25 years	$15 \leq n \leq 25$
Parameters a and b in the logistic function are configured based on the full membership of speed in the medium speed fuzzy set, that is, $n = 20$ and $\psi = 10$	$a = 1.01$ $b = 98.99$ $c = \psi/n = 0.5$
The rule-firing threshold for a cell to have strong propensity for development	$\mu_0 = 0.5$
Neighbourhood support	$\mu_{\max S_{\text{urban}}}(\Omega_{x_{ij}}) > \mu_{S_{\text{urban}}}(x_{ij})$

With the aforementioned configurations, the six primary transition rules as described in Chapter 3, Section 3.3.3.1, are implemented as a number of IF–THEN statements, as shown in Table 5.2. The initial settings of the model and its relevant parameters can be adjusted during the model calibration process.

5.1.3.2 Constrained Development by Secondary Rules

Apart from the primary transition rules, a number of secondary transition rules reflecting the impact of topographical conditions, transportation network, as well as urban planning schemes have also been applied to model the process of the urban development of Sydney.

5.1.3.2.1 Topographical Conditions

As discussed in Chapter 4, historically, the urban expansion of Sydney was largely constrained by topographical conditions (Haworth 2003). This topographical constraint is primarily reflected by the slope of land and terrain features. Although the slope of land is not a determining factor, it can speed up or slow down the process of urban development to some extent. Therefore, the slope factor was introduced into the cellular automata model as a constraint on the urban development of Sydney. To implement this constraint, it was assumed that the transition rules implemented in the model on urban natural growth were based on a flat landscape, that is, a slope of 0–2 degrees. If the slope of a cell is moderate, a slow pattern of development might occur;

TABLE 5.2
The Primary Transition Rules

For cells located in an area excluded from urban development (e.g., water bodies, national parks, natural reserves, etc.)
  No development occurs to the cell

For cells that are fully developed into an urban state, that is, $\mu_{S_{urban}}(X_{ij}) = 1$
  No further development occurs to the cell

For cells that are in a partly urban state, that is, $0 < \mu_{S_{urban}}(X_{ij}) < 1$

IF	$\mu_{S_{urban}}(X_{ij}) \geq \mu_0$ AND $\mu_{maxS_{urban}}(\Omega_{x_{ij}}) > \mu_{S_{urban}}(X_{ij})$	Primary Rule 1 (PR1)
THEN	the cell continues to develop at a *medium* speed	
ELSE IF	$\mu_{S_{urban}}(X_{ij}) \geq \mu_0$ AND $\mu_0 \leq \mu_{maxS_{urban}}(\Omega_{x_{ij}}) < \mu_{S_{urban}}(X_{ij})$	Primary Rule 2 (PR2)
THEN	the cell continues to develop at a *slow* speed	
ELSE IF	$0 < \mu_{S_{urban}}(X_{ij}) < \mu_0$ AND $\mu_{maxS_{urban}}(\Omega_{x_{ij}}) > \mu_{S_{urban}}(X_{ij})$	Primary Rule 3 (PR3)
THEN	the cell continues to develop at a *slow* speed	
ELSE IF	$0 < \mu_{S_{urban}}(X_{ij}) \leq \mu_0$ AND $0 < \mu_{maxS_{urban}}(\Omega_{x_{ij}}) < \mu_{S_{urban}}(X_{ij})$	Primary Rule 4 (PR4)
THEN	the cell continues to develop at a *very slow* speed	

For cells that are in a non-urban state, that is, $\mu_{S_{urban}}(X_{ij}) = 0$

IF	$\mu_{maxS_{urban}}(\Omega_{x_{ij}}) \approx 1$	Primary Rule 5 (PR5)
THEN	the cell may start to develop at a *slow* speed	
ELSE IF	$\mu_0 \leq \mu_{maxS_{urban}}(\Omega_{x_{ij}}) < 1$	Primary Rule 6 (PR6)
THEN	the cell may start to develop at a *very slow* speed	

For all other cells, no development will occur to the cells

TABLE 5.3

The Slope Constraint on Urban Development

Primary rules	Flat (slope = 0~2°)	Moderate (slope =3~5°)	Steep (slope = 6~10°)	Very steep (slope > 10°)
	Slope constraint			
PR1	Normal development	Slow development	Very slow development	Extremely slow development
PR2, PR3, or PR5	Slow development	Very slow development	Extremely slow development	No development
PR4 or PR6	Very slow development	Extremely slow development	No development	No development

for a cell with a steep or even very steep slope, the development speed will be very slow. Therefore, four linguistic variables describing the slope of the landscape were proposed: *flat, moderate, steep,* and *very steep.* The initial configuration of these variables and their impacts on the primary transition rules are listed in Table 5.3. However, the initial values of these linguistic variables may be modified during the model calibration process.

The topographical conditions of a region may also affect urban development in positive ways. In particular, the relatively high terrain and beautiful natural scenery along the eastern coast have caused the development of non-urban land into urban areas in Metropolitan Sydney. This factor is built into the model as a favourable driver to urban development. The rule-firing conditions of this factor are listed in Table 5.4.

TABLE 5.4

Rules on Terrain and Coastal Proximity Attractions Applicable to Non-Urban Cells

Primary conditions	Slope condition	Terrain condition	Proximity to coast	Other conditions	Type of development that can occur to the cell
		Secondary conditions			
PR6	Flat to moderate	With coastal view	Within proximity to coast	Nil	Slow development
Nil	Flat to moderate	With coastal view	Within proximity to coast	Nil	Very slow development
PR6	Steep	With coastal view	Within proximity to coast	Nil	Very slow development
Nil	Steep	With coastal view	Within proximity to coast	Pass a random selection	Extremely slow development

5.1.3.2.2   The Transportation Network

Transportation is an important factor in accelerating urban development and attracting new development. For Sydney, the early development of the region was closely tied to the development of its public transportation system (New South Wales Department of Planning 1993b). From the mid-19th century, the urban area of Sydney expanded outward along the radial rail lines. Most urban constructions were conducted along the extensive network of tramlines in the late 19th century. From the mid-20th century, the electrification of railways and the construction of the city's underground railway greatly improved the overall capacity of the existing network and encouraged commuting from greater distances. Moreover, a high level of car usage has enabled a greater extension of suburban settlements in the less accessible areas of Sydney. Hence, transportation is an important factor that needs to be introduced into the model.

A good transportation network increases the accessibility of land. Consequently, areas with good accessibility are more easily selected for urban development. As different transportation modes and different standards of transportation have different strengths of impact or potential to attract new development and/or accelerate existing development, a transportation density index (TDI) was designed to measure the spatial accessibility of a cell in the urban development content. This TDI consists of two components: a line density index for roads and railways, and a point density index for railway stations.

The point density index for railway stations is included in the TDI based on the understanding that areas close to railway stations are more accessible than others and therefore, are considered important to lead to further or new development, especially when a railway station is close to other urban centres or localities. This density index is applied to measure the density of railway stations, magnified by different weights, which are within a specified neighbourhood of each output raster cell. The weights assigned to railway stations are based on the distance from a station to the nearest urban centre. The closer a railway station to the centre, the higher the weight. Table 5.5 displays the different weights applied to different railway stations within the Sydney Metropolitan Area; these weights can be modified during the model calibration process.

TABLE 5.5
The Initial Assignment of Weights to Railway Stations

Direct distance of the railway station to the nearest urban centre	Weight
Within 1 km	4
1–3 km	3
3–5 km	2
5–10 km	1
More than 10 km	0

The line density index for roads and railways is computed using a simple line density method (Silverman 1986). This index is chosen to measure the density of linear features that fall within a certain identified neighbourhood of each output cell. Similar to the point density index, a weighting system was proposed to quantify the different modes and classes of transports. For instance, a local or residential road will carry less weight in this index compared to a secondary road or a railway line. A dual carriageway such as the freeway or a national highway with links to the local transportation network will carry a higher weight than all other transport modes. The initial assignment of weights to the various transport modes and categories is listed in Table 5.6. These weights will also be modified during the model calibration process.

Both the point and line density indexes can be computed using Environmental Systems Research Institute's (ESRI) ArcGIS Spatial Analyst extension (ESRI 2004a). A higher TDI suggests a higher level of accessibility, resulting in a stronger impact of transport on urban development at the locality.

To implement the rules of transportation support on urban development into the model, three fuzzy linguistic terms were used to represent the strength of the transportation support a cell can receive. If a cell has a very high TDI, it is entitled to have a *strong transportation support* for urban development. If the TDI of a cell is not very high but is strong enough to expedite existing development or attract new development, it is entitled to have a *weak transportation support*. If the TDI of a cell is very low, then the cell is entitled to have *no transportation support*.

With a strong transportation support, a partly urban cell might be further developed very quickly, or a non-urban cell may be selected for development at a speed faster than one without strong transportation support. On the other hand, development can also be accelerated or new development initiated in areas having a weak transportation support, although this development may not be as quick as the one that has a strong transportation support. The impact of transportation on the speed of urban development is summarised in Table 5.7. The fuzzy thresholds of TDI quantifying the strength of transportation support can be configured and adjusted through model calibration as well.

TABLE 5.6
The Initial Assignment of Weights to Different Transportation Modes

Transport type	Weight
Dual carriageways	4
Multiple track railways with standard gauge or principal roads	3
Single track railways with standard gauge or secondary roads	2
Sealed minor roads (local or residential roads)	1
Unsealed minor roads or tracks	0

TABLE 5.7

The Secondary Transition Rules on Transportation Support

Primary rules	Secondary rules	Transportation support		
		No transportation support	Weak transportation support	Strong transportation support
PR1		Normal development	Fast development	Very fast development
PR2, PR3, or PR5		Slow development	Normal development	Fast development
PR4 or PR6		Very slow development	Slow development	Normal development

5.1.3.2.3   Urban Planning Schemes

Apart from the most recent Metropolitan Strategic Plan, which was released in 2005, there have been three metropolitan planning schemes that may have affected the urban development in Sydney since 1976. They are the Sydney Region Outline Plan (New South Wales State Planning Authority 1968), Sydney into its Third Century (New South Wales Department of Environment and Planning 1988), and Cities for the 21st Century (New South Wales Department of Planning 1995). In each of these plans, a number of areas were identified for urban development within a certain time frame. However, to what extent the actual urban development of Sydney followed the blueprints of these planning schemes needs to be evaluated.

The impact of urban planning schemes on actual urban development can be reflected in two aspects. One is to act as an accelerating factor to promote more new development in areas planned or proposed for development, and the other is as a global factor to constrain or accelerate the overall development of cells within a region. Introducing urban planning rules will change the speed of development or the transition of cells from one state to another in areas affected by the planning schemes within the specified time frame, either by speeding up or slowing down the process of such development, based on the nature of the planning schemes.

For instance, in the Sydney Region Outline Plan proposed in 1968, a number of growth corridors or centres to the north, west, south-west, and north-west of the metropolitan area were identified. In order to represent the impact of these planning initiatives, a secondary transition rule was added to the model in accordance with the time frame of the plan. Those corridors and centres planned for development were affected by this rule as an accelerating factor either by releasing the constraint or reinforcing the support of other secondary factors such as slope or transportation, and resulted in more opportunity for new development in those proposed areas. However, the strength of this planning factor on the cell's development needs to be calibrated, which is realised through the setting of a strength parameter. The value of this parameter is to be configured and calibrated during the model calibration process.

In addition to proposing areas for urban development, the Outline Plan was based on high population growth, which may lead to faster development in all areas that satisfy primary development conditions. This planning policy is reflected in the model by modifying the overall speed of development (i.e., the value of n in the model), which also results in the release of other constraints on the development of the cells.

Similarly, transition rules were also proposed reflecting the impacts of the 1988 Sydney into its Third Century plan and the 1995 Cities for the 21st Century plan on urban development in two ways: by promoting more development in areas identified for further development, and by accelerating or reducing the speed of development in other areas. Whether the planning scheme is acting as an accelerating factor or a constraining factor and when the factor is introduced into the model are dependent on the planning scheme itself. For instance, both the 1988 and 1995 urban planning schemes emphasised the significance of urban consolidation and redevelopment in established areas. Hence, these planning policies should be reflected in the model as constraining factors that reduce the speed of urban expansion within the specified time frame of the respective planning scheme.

Apart from the aforementioned three planning schemes, the newly released metropolitan plan City of Cities is introduced into the model only when the model is applied to project the future urban scenarios of Sydney. These will be discussed in Chapter 6.

5.1.3.3 Flexibility in Rule Implementation

So far, this chapter has demonstrated a flexible way in which various factors that may impact urban development in Sydney can be introduced into the cellular automata model through either primary or secondary transition rules. Four key factors, such as the self-propensity for development and neighbourhood support, slope constraints and terrain or coastal proximity attractions, transportation support, and urban planning programmes, have been introduced into the model to simulate Sydney's urban development from 1976 to 2006. Other factors, such as the urban infrastructure (drainage and sewerage systems), income level, and accessibility to community services can also contribute to urban development in Sydney. These factors may either accelerate or slow down the speed of the development process, which can be introduced into the model as well provided that a good understanding of the impact of such factors on urban development is maintained and quality data representing such factors are available.

According to the general principles on rule inferencing outlined in Chapter 3, Section 3.3.3.3, the aggregate impacts of the accelerating or constraining factors on the model's behaviour are as follows:

1. If there is one accelerating factor within the neighbourhood of the cell in question, the speed of urban development of that cell will be upgraded one step higher in the "speed" fuzzy set.
2. If there is one constraining factor within the neighbourhood of the cell in question, the speed of urban development of the cell will be downgraded one step lower in the "speed" fuzzy set.

3. If there is more than one such factor, the speed will be upgraded or down-graded two steps up or down.
4. The existence of one accelerator and one constraint will cancel the effect of both factors; hence, the speed of development will stay unchanged.

The flexibility of the model enables it to function not only as an analytical tool to understand the forces driving the process of urban development but also as a planning tool to experiment with various planning proposals and generate different "what if" scenarios in the planning practice.

5.1.4 THE TEMPORAL DIMENSION

As discussed in Chapter 2, Section 2.3.5, many cellular automata-based urban models were not configured temporally. Those models were typically configured by starting the model from a certain time at which spatial data sets were available, and running the model for a number of iterations until the simulated results fit with the reference data at the ending point of time. Therefore, the temporal interval of the simulation time was not confirmed. This type of model configuration cannot be applied in simulating the temporal process of urban growth.

The fuzzy constrained cellular automata model developed in this book is configured temporally when applying it to simulate the process of urban growth in Sydney. For the calibration of the model, the starting date was set to 1976, and the ending date to 2006. Each temporal advance of the model represents one year in the urban growth context. Data illustrating the urban extent within the simulation period at every 10-year interval were used to calibrate the model temporally. The methodology for calibration is elaborated in the following section.

5.2 MODEL CALIBRATION

Models are simplified representations of real systems. As a result of this simplicity, a model can never completely reproduce the structure or process of the modelled system (Norlén 1975). The model leaves part of the reality behind. This is both its power and its weakness. The power lies in the clarity of the essentials and the manipulative nature of the symbols; the weakness lies in a small but necessary degree of invalidity (Kilbridge, O'Block, and Teplitz 1970). Therefore, testing or verifying how well a model matches its specifications, and minimising or controlling the model's degree of invalidity, become two critical tasks in model construction (Naylor and Finger 1967). As Reichenbach states, "verifiability is a necessary constituent of the theory of meaning. A sentence the truth of which cannot be determined from possible observations is meaningless" (Reichenbach 1951: 256–7). Similarly, a model whose results are not validated is also of limited value.

5.2.1 MODEL CALIBRATION PRINCIPLES

Calibration is the process of verifying that a measuring object is performing within its designated accuracy. It is often used to estimate a model's parameters that provide the best fit to an observed set of data (Bell, Dean, and Blake 2000: 574). The validity of a simulation model is determined by "the accuracy of its predictions" (Kilbridge, O'Block, and Teplitz 1970: 33). However, comparing the model's forecast with the eventual reality is not a true validity test. Conditions not included in the original model may eventuate to make its forecasts inaccurate, even though the model's first conception is valid. On the other hand, a poorly conceived model may give an accurate forecast by accident, as it may have "done the right thing for the wrong reason" (Kilbridge, O'Block, and Teplitz 1970: 33).

A number of statistical measures and techniques have been developed for testing the degree of conformity or the "goodness-of-fit" of the simulated time series to the observed data (Naylor and Finger 1967: 97). Among them, factor analysis, regression, and principal component analysis (PCA) have been widely used in urban modelling since the 1960s (Theobald and Hobbs 1998; Naylor and Finger 1967; Cohen and Cyert 1961). More recently, measures such as multicriteria evaluation and analytical hierarchy process (AHP) approaches have also been applied to estimate the setting of parameters in urban models, based on the cellular automata approach (Wu and Webster 1998; Wu 1998c).

Similar to other simulation models, the model developed in this book was tested using a retrospective prediction approach, that is, predicting the known past and present using a historical data set. Based on the principles of cellular automata, complex states of cells can be modelled by simple rules provided that the base state of a system and the transition rules have been correctly identified. This philosophic frame is both helpful and discouraging. It is helpful because it suggests that one can model complex systems with simple rules. Such a philosophic frame is discouraging when the behaviour of a cellular automaton relies absolutely on the accurate identification of initial states and the correct setting of transition rules.

As urban development is a continuous process that is affected by many factors, it is a fuzzy process both spatially and temporally, making the identification of urban states, and hence, the setting of transition rules that control urban development difficult. In this case, it is necessary to start modelling by choosing the minimum number of rules and making them as simple as possible, so that one may get the best chance to identify the transition of states of cells accurately within the complexity of such rules. The trade-off between model complexity and the fidelity of real-world representation is necessary for setting realistic and transparent rules in cellular automata-based urban modelling. With this understanding, the following principles were set up for the calibration of the cellular automata model of the urban development of Sydney:

Principle I	Start the model with the simplest case, that is, using the simplest rules possible, and assuming that the same rules apply evenly throughout in space and time
Principle II	Change the parameters of the rules within each transition
Principle III	Change the balance between the rules
Principle IV	Increase the sophistication of some of the rules
Principle V	Introduce new rules if necessary

These five principles were applied consecutively step by step. The second principle was applied only when the first principle had proved to be inadequate in terms of model calibration, and the third, fourth, and fifth principles were applied only when both the first and the second principles were inadequate. Although this approach provides an interactive way of calibrating the model, one should be aware that, within a limited number of calibrations, the model might not generate peak performance because of the setting of the parameters. Therefore, it is important that the calibration of the model be undertaken systematically and adequately.

The adequacy of the model's results was evaluated by comparing them with actual urban development in three consecutive ways, that is, visual calibration, statistical calibration, and calibration over time. Visual calibration was useful in the initial phase of modelling to establish the rough parameter settings. This calibration was necessary, especially for verifying that the model was replicating the spatial pattern and extent of historical development. It cannot be replaced by statistical calibration alone (Clarke and Gaydos 1998). Once the initial visual calibration had been verified, a further step for testing the accuracy of the simulated results with the observed data was carried out through a simulation accuracy assessment.

5.2.2 SIMULATION ACCURACY ASSESSMENT

One important aspect of model calibration is to verify the model's outcomes and evaluate the goodness-of-fit of those outcomes with the real-world system it modelled. One way of evaluating such goodness-of-fit is through an accuracy assessment. Accuracy assessment can be defined as the task of comparing two maps: one generated by the model (data to be assessed), and the other based on the ground truth (the reference data). The reference data are assumed to be accurate and forms the standard for comparison. A simple evaluation compares the two maps with respect to the areas assigned to each category. This assessment only considers the overall areas of agreement in each category of the two maps. It does not consider any agreement between the two maps at specific locations. Therefore, it is termed a *non-site-specific accuracy*, which may give misleading results (Campbell 1996).

Another form of accuracy, *site-specific accuracy*, is based on the detailed assessment of agreement between the two maps at specific locations (Campbell 1996). The assessment of site-specific accuracy is commonly conducted by preparing an error matrix, which is discussed in the next section.

5.2.2.1 The Error Matrix Approach

The error matrix approach is a common means of expressing the accuracy of land-cover classification for remotely sensed data (Jensen 1996; Lillesand and Kiefer 1994; Story and Congalton 1986). This matrix summarises data from two different sources (one from reference data representing the ground truth, and the other from classified image data), and compares the relationship between the two on a site-by-site basis. An error matrix is a square array, the rows and columns of which represent the number of categories whose classification accuracies are being assessed. Typically, the columns of an error matrix represent the reference data, whereas rows indicate the classified image data. The terms used in an error matrix and their calculations are defined in Table 5.8.

TABLE 5.8
Definitions of Terms Used in an Error Matrix and Their Computations

Terms	Definitions	Computations
Producer's accuracy	Probability of reference cells being correctly categorised in the classification data. This measures the omission error.	Number of cells on the major diagonal divided by the column total of each category
User's accuracy	Probability that cells in the classification data actually belong to the same category as in the reference data. This measures the commission error.	Number of cells on the major diagonal divided by the row total of each category
Omission error	Cells that are excluded (omitted) from the categories that they belong to in the reference data	Total of the off-diagonal column cells divided by the column total of each category
Commission error	Cells that are included (committed) in the categories that they do not belong to in the reference data	Total of the off-diagonal row cells divided by the row total of each category
Overall accuracy	A measurement of the overall proportion of correctly categorised cells in relation to the total number of cells under assessment	Total number of cells along the major diagonal of the matrix divided by the total number of all the cells

In an error matrix, cells that are categorised in agreement with their reference data are located along the major diagonal of the error matrix, running from the upper left to the lower right. All off-diagonal cells represent errors of commission or omission (Lillesand and Kiefer 1994). By examining the relationship between the commission and omission errors, users gain insight into the reliability of categories of output data generated by the classification, and producers learn about the performance of the process that generates the data. Table 5.9 is a sample error matrix illustrating a cell-by-cell comparison of satellite-based land-use classification and ground reference classes.

Table 5.9 shows that the producer's accuracy for each category of land use ranges from 63.3% to 90.4%; the user's accuracy also ranges from 75.6% to 90.2%. For instance, the producer's accuracy for water body is 232/292 or 79.5%. This informs the producer who prepared the classification that, of the actual cells considered as water bodies in the ground reference data, 79.5% were correctly classified in the satellite-based land classification. For the same water body category in Table 5.9, the user's accuracy is 232/307 or 75.6%. This informs the user of the classified data that, of all the cells classified under the water body category, 75.6% actually corresponds to the water body on the ground. In other words, the omission error permits identification of water bodies mislabelled as other categories, and the commission error permits identification of the cells erroneously labelled as water body. The overall accuracy shows that 81.1% of all cells under assessment were correctly categorised in this classification. The error matrix is very effective in representing classification accuracy as it clearly describes the overall accuracy and accuracy of each category along with its commission and omission errors.

TABLE 5.9

An Error Matrix Illustrating a Cell-by-Cell Comparison of Satellite-Based Land-Use Classification and Ground Reference Classes

		Ground reference classes					
		Water body	Barren area	Agricultural land	Dense vegetation	Shrub/ grass	Row total
Satellite-based land classes	Water body	**232**	45	14	11	5	307
	Barren area	33	**465**	26	40	23	587
	Agricultural land	0	5	**148**	2	9	164
	Dense vegetation	22	35	24	**587**	33	701
	Shrub or grass	5	14	6	9	**121**	155
	Column total	292	564	218	649	191	1914

Producer's accuracy			Omission error
Water body	= 232/292	= 79.5%	20.5%
Barren area	= 465/564	= 82.4%	17.6%
Agricultural land	= 148/218	= 67.9%	32.1%
Dense vegetation	= 587/649	= 90.4%	9.6%
Shrub/grass	= 121/191	= 63.3%	36.7%

User's accuracy			Commission error
Water body	= 232/307	= 75.6%	24.4%
Barren area	= 465/587	= 79.2%	20.8%
Agricultural land	= 148/164	= 90.2%	9.8%
Dense vegetation	= 587/701	= 83.7%	16.3%
Shrub/grass	= 121/155	= 78.1%	21.9%
Overall accuracy	**= 1553/1914**	**= 81.1%**	

5.2.2.2 A Modified Error Matrix Approach

Although the error matrix approach is effective in assessing the accuracy of land-use and land-cover classification of remotely sensed images, a limitation has been identified in using this matrix to assess simulation accuracies. This is because the error matrix approach is based on a cell-by-cell comparison of two maps. The comparison does not necessarily capture the qualitative similarities, that is, the similarity of patterns between the maps (Power, Simms, and White 2001). For a simulation model of urban growth, it is the pattern of cells that has functional significance in the urban development context. Therefore, it is important to assess whether the spatial patterns of cells in the simulation output match the corresponding patterns in the reference map. However, as the error matrix compares two maps on a cell-by-cell basis, there could be large omission or commission errors between the results of the model and its actual reference data, even though the patterns of cells are similar (Figure 5.2). This may result in low individual and overall accuracies (Table 5.10).

Although various approaches have been developed for pattern recognition (Friedman and Kandel 1999; Theodoridis and Koutroumbas 1999; Webb 1999; Devroye, Gyorfi, and Lugosi 1996; Gose, Johnsonbaugh, and Jost 1996; Fukunaga 1990), their application

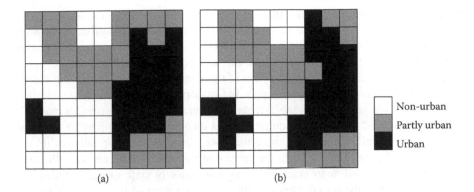

FIGURE 5.2 Comparing a simulated result with its reference data. The two maps show that the patterns of the state of cells are similar although they generate low simulation accuracies when compared on a cell-by-cell basis: (a) Reference data and (b) results generated by the model.

in assessing the accuracy of a simulation model was not identified, with the exception of the hierarchical fuzzy pattern-matching approach developed by Power, Simms, and White (2001) to compare land-use maps.

This chapter presents a modified error matrix approach to assess the accuracy of the simulation model of urban development. Recall that the state of cells in the cellular automata model developed in Chapter 3 was defined using a fuzzy membership grade; the value of each cell in the urban fuzzy set ranges from 0 (non-urban state) to 1 (fully urban state), representing an indefinite cell state in the urban fuzzy set. After the operation of the model, the value of cells generated by the model also ranges from 0 to 1. These values were defuzzified into three categories using a simple linear

TABLE 5.10
The Error Matrix of the Two Maps Shown in Figure 5.2

		Reference data[a]			
		Non-urban	**Partly urban**	**Urban**	**Row total**
Simulated results[a]	Non-urban	24	5	2	31
	Partly urban	1	21	3	25
	Urban	3	4	18	25
	Column total	28	30	23	81
Producer's accuracy				**Omission error**	
	Non-urban	= 24/28	= 85.7%	4.3%	
	Partly urban	= 21/30	= 70.0%	30.0%	
	Urban	= 18/23	= 78.3%	21.7%	
User's accuracy				**Commission error**	
	Non-urban	= 24/31	= 77.4%	22.6%	
	Partly urban	= 21/25	= 84.0%	16.0%	
	Urban	= 18/25	= 72.0%	28.0%	
Overall accuracy		**= 63/81**	**= 77.8%**		

Note: [a] Data shown as total number of cells in each category.

defuzzification function. These three categories were termed non-urban, partly urban, and urban. Cells with a membership grade of 0 were non-urban, cells with a membership grade of 1 were urban, and all other cells whose membership grades ranged between 0 and 1 exclusively were categorised as partly urban cells.

In order to enhance the pattern effect of the simulated results, a moving kernel of 3 × 3 cells was designed to compute the majority state of cells (the state that appears most frequent) within the extent of the kernel. Subsequently, this majority state was assigned to the corresponding cell as its new membership value to define the state of the cell. If there was more than one majority state for cells within the extent of the kernel, the one with the same state as the output cell was allocated to the processing cell as its state. By using the majority state of a cell within the neighbourhood of the moving kernel rather than the state of the cell itself in preparing the error matrix, the similarity of patterns between the model's results and the reference data can be more clearly identified. Figure 5.3 shows the majority state of the cells in Figure 5.2b.

To evaluate the accuracy of the model's results, the new state of cells represented by the majority state of the neighbouring cells within the 3 × 3 moving kernel was compared with the reference data. For the example given in Figure 5.2, the modified error matrix computed by the new states of the output data is presented in Table 5.11. It shows that both the producer's and the user's accuracies in each category increased, and the overall accuracy of the result increased from 77.8 to 82.7%. This is because the modified error matrix used the majority state of cells, thus enhancing the recognition of the pattern of cells of the model's output.

5.2.2.3 Kappa Coefficient Analysis

Inspection of the error matrix reveals the overall nature of the errors in the classification or simulation results. Both the individual and overall accuracies use the main diagonal cells of the matrix to measure the percentage of correctly categorised cells, and as such, they are relatively simple and intuitive measure of agreement (Ma and Redmond 1995). In addition, the percentage of correctly categorised cells does not take into account the proportion of agreement between the two data sets that are due to chance matching. This percentage tends to overestimate the accuracies (Rosenfield 1986; Rosenfield and Fitzpatrick-Lins 1986; Congalton and Mead 1983; Congalton, Oderwald, and Mead 1983).

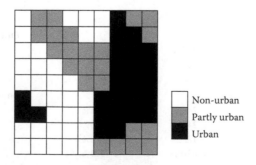

Non-urban
Partly urban
Urban

FIGURE 5.3 The majority state of cells in Figure 5.2b within a 3 × 3 kernel.

TABLE 5.11

The Modified Error Matrix of the Two Maps: One Shown in Figure 5.2a and One in Figure 5.3

		Reference data[a]			
		Non-urban	Partly urban	Urban	Row total
Simulated results[a]	Non-urban	27	6	0	33
	Partly urban	1	20	3	24
	Urban	0	4	20	24
	Column total	28	30	23	81
Producer's accuracy				**Omission error**	
	Non-urban	= 27/28	= 96.4%	3.6%	
	Partly urban	= 20/30	= 66.7%	33.3%	
	Urban	= 20/23	= 87.0%	13.0%	
User's accuracy				**Commission error**	
	Non-urban	= 27/33	= 81.8%	18.2%	
	Partly urban	= 20/24	= 83.3%	16.7%	
	Urban	= 20/24	= 83.3%	16.7%	
Overall accuracy		**= 67/81**	**= 82.7%**		

Note: [a] Data shown as total number of cells in each category.

Chrisman (1980); Congalton and Mead (1983); and Congalton, Oderwald, and Mead (1983) proposed an application of the Kappa analysis, as described by Bishop, Fienber, and Holland (1975) and Cohen (1960), as a means of improving the interpretation of the error matrix. Kappa analysis yields a K_{hat} coefficient that measures the difference between the observed agreement of the two maps and the agreement that might be attained by chance matching (Campbell 1996). The K_{hat} coefficient is computed as follows:

$$K_{hat} = \frac{N \sum_{i=1}^{r} x_{ii} - \sum_{i=1}^{r} (x_{i+} \times x_{+i})}{N^2 - \sum_{i=1}^{r} (x_{i+} \times x_{+i})} \qquad (5.1)$$

where
   r is the number of categories in the error matrix
   x_{ii} is the number of cells in row i and column i
   x_{i+} is the total number of cells in row i (shown as the row total in the matrix)
   x_{+i} is the total number of cells in column i (shown as the column total in the matrix)
   N is the total number of observations included in the matrix

Using data in Table 5.11, Table 5.12 demonstrates the computation of the K_{hat} coefficient, which is only 74.0% and is significantly lower than the overall accuracy of 82.7%. This is because the two measures incorporate different information. The overall accuracy data incorporates only the major diagonal cells and excludes the omission and commission errors, whereas the K_{hat} coefficient also incorporates the off-diagonal cells as a product of the row and column marginals.

TABLE 5.12

Computation of K_{hat} of the Two Maps: One Shown in Figure 5.2a and One in Figure 5.3

$$K_{hat} = \frac{N \sum_{i=1}^{r} x_{ii} - \sum_{i=1}^{r}(x_{i+} \times x_{+i})}{N^2 - \sum_{i=1}^{r}(x_{i+} \times x_{+i})}$$

$r = 3$ (i.e., non-urban, partly urban, and urban)

$N = 81$ (i.e., 9 rows $\times$ 9 columns)

$\sum_{i=1}^{r} x_{ii} = (27 + 20 + 20) = 67$

$\sum_{i=1}^{r} (x_{i+} \times x_{+i}) = (28 \times 33 + 30 \times 24 + 23 \times 24) = 2184$

therefore, $K_{hat} = \dfrac{81 \times 67 - 2196}{81^2 - 2196} = 74.0\%$

The K_{hat} coefficient adjusts the percentage of correctly categorised cells by subtracting the estimated contribution of chance agreement. Thus, the value of K_{hat} in Table 5.12 can be interpreted to mean that the simulation achieved an accuracy that is 74.0% better than what would be expected from chance assignment of cells to categories. As the overall accuracy approaches 100%, and the contribution of chance agreement approaches 0%, the value of K_{hat} approaches 100%, indicating a perfect effectiveness of the simulation results. On the other hand, as the effect of chance agreement increases and the percentage of correctly categorised cells decreases, K_{hat} assumes a negative value (Campbell 1996).

5.3   MODEL IMPLEMENTATION IN GIS

5.3.1   Cellular Automata Modelling and GIS

Cellular automata models can be implemented within many types of software (Batty 1997). For cellular automata models of urban development, programs have been implemented both inside and outside a GIS environment. In regard to using a GIS environment, Itami and Clark (1992) and Itami (1988) developed their models within a raster GIS of Idrisi and MAP II. Wu (1998a,b,c) and Li and Yeh (2001) implemented their cellular automata models using ARC/INFO's Arc Macro Language (AML). More recently, Stevens and Dragicevic (2007) developed their iCity model as a fully integrated extension of ArcGIS 9. These programs took advantage of the wide range of spatial analysis functionalities, the graphic display and visualisation capabilities, as well as the friendly user interface of a GIS. Moreover, as most applications of cellular automata models need to consider the use of various data sources including historical land-use maps, census data, as well as satellite imagery, data processing and integration from different formats can be more easily implemented using an industry standard GIS software program.

With these advantages, geographical models developed within a GIS environment are preferable in many applications.

Some cellular automata models were developed on custom-built software outside a GIS environment, although most were loosely coupled with a GIS for data manipulation and visualisation. This kind of model is exemplified by Batty, Xie, and Sun (1999); Batty (1998); and Clarke, Hoppen, and Gaydos (1997). Such models were usually designed to meet the needs of users, and they did not rely on any commercial GIS software to support their operation. However, the development of these models can be technically more challenging as it usually requires a high level of specialisation in software development (Batty, Xie, and Sun 1999; White, Engelen, and Uljee 1997).

In this book, a strong coupling approach was adopted to implement the fuzzy constrained cellular automata model of urban growth within a GIS environment. The model was initially programmed using ARC/INFO's AML in a GRID environment. This was later converted to an ArcGIS extension by using Visual Basic for Applications (VBA) within ESRI's ArcGIS 9 development environment.

There were a number of reasons for using a strong coupling strategy. One was that data were initially processed and stored in ArcGIS and were converted into grid files, and the simulated outputs of the model were also stored as raster grid files in ArcGIS. No data conversion was necessary between the GIS and the model, which saved time on data communication. This feature was especially advantageous during the model calibration process when simulation results were compared and fitted with data illustrating actual urban development. Another useful feature of the model was its spatial visualisation capability. As all input data and output results were stored and processed within the same GIS environment, the results could be easily visualised spatially using the data display and visualisation capabilities of ArcGIS. Moreover, the design of a friendly graphic user interface (GUI) makes it possible for users to modify and calibrate the model vigorously. The accuracies of the model's simulation outcomes in comparison with actual urban scenarios could also be computed and presented conveniently to the user for evaluation and model calibration.

5.3.2   The ArcGIS Approach

ArcGIS 9 is a desktop product of the Environmental Science Research Institute (ESRI). This is a well-documented GIS program that has been widely used by many in the private and public sectors in the geospatial community. It is a powerful program for users to perform geospatial data manipulation, processing, analysis, and cartographic visualisation. Apart from the many tools and functions that ArcGIS provides, such as data management, cartographic presentation, spatial analysis, geocoding, geoprocessing, and so on, it also provides a well-documented software development kit (SDK), which allows model developers to programmatically access ArcGIS to automate repetitive tasks and construct spatial-based models to extend its functionalities. This is achieved through third-party component object model (COM)-compliant programming languages such as Visual Basic 6, Visual Basic, C, Visual C++, Java, .NET, or Python (ESRI 2004b).

Unlike other COM-compliant languages, VBA is not a stand-alone application; it is embedded within ArcGIS applications, including ArcMap™ and ArcCatalog™. Therefore, once ArcGIS is available, there is no need to purchase any other software for programming.

VBA is a powerful Windows application development tool. It is an implementation of Visual Basic (VB), a common and popular programming language. VBA enables developers to write code to build customer solutions, automate workflows, and extend the functionality of an application (ESRI 2004b). A Visual Basic Editor (VBE) was embedded in ArcGIS, which allows users to write VB macros to customise existing or create new user interfaces and develop custom GIS applications. Moreover, as VB is an object-oriented development tool, developers can make use of the many ArcObject libraries provided by ESRI within the VBA environment.

5.3.3 GRAPHIC USER INTERFACE DESIGN

The fuzzy constrained urban cellular automata model was developed and implemented as an ArcGIS extension using the VBA development tool. This ArcGIS extension is called *FuzzyUrbanCA*, which can be launched from within the ArcMap application.

A simple GUI was designed. This friendly user interface enables users to configure the model, including the setting of cell size and neighbourhood size and the selection of primary and secondary transition rules. Through this GUI, users can also execute and calibrate the model vigorously during the simulation process. Figure 5.4 shows a screenshot of the GUI.

With the GUI, users first need to decide the cell size and the neighbourhood size that the model will be operated on. The default cell size is 250 m; this is the only cell size available for model calibration in the current version of the model.

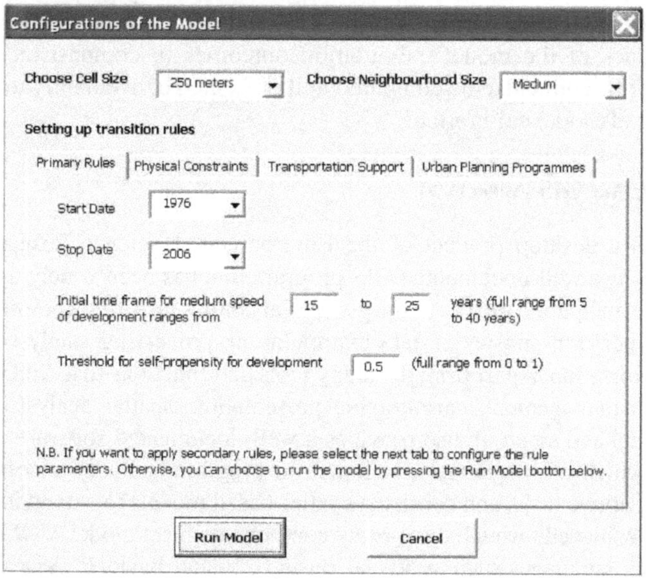

Configurations of the Model

Choose Cell Size [250 meters ▾]   Choose Neighbourhood Size [Medium ▾]

Setting up transition rules

Primary Rules | Physical Constraints | Transportation Support | Urban Planning Programmes

Start Date [1976 ▾]

Stop Date [2006 ▾]

Initial time frame for medium speed of development ranges from [15] to [25] years (full range from 5 to 40 years)

Threshold for self-propensity for development [0.5] (full range from 0 to 1)

N.B. If you want to apply secondary rules, please select the next tab to configure the rule parameters. Otherwise, you can choose to run the model by pressing the Run Model botton below.

[Run Model]   [Cancel]

FIGURE 5.4 A snapshot showing the graphical user interface (GUI) of the model.

Other cell sizes, such as 100m and 500m, can be added into the model to evaluate the sensitivity of the model to cell scales. This task will be achieved at a later stage. There are three options for neighbourhood size, which ranges from a small neighbourhood with a radius of one cell to a very large neighbourhood with a radius of four cells. The default setting is a medium neighbourhood with a radius of two cells. Users can change the neighbourhood size to a small, a large, or a very large one to evaluate the impact of neighbourhood scale on simulation behaviours and outcomes.

For the configuration of transition rules, users can choose to apply only the primary transition rules, or they can also select other secondary transition rules to evaluate the impact of different rule settings on the model's simulation outcomes. With the flexibility in implementing secondary transition rules, additional rules can also be added to the model, provided that a good understanding of the rule or the associated factor on urban development is achieved and relevant data are available to reflect the impact of this factor on urban development. However, the incorporation of new transition rules can only be achieved at the programmer level, not at the user level. Once a user has selected relevant data and rules for the initial configuration of the model, the model can be executed by pressing the "Run Model" button at the GUI (Figure 5.4).

5.3.4 MODEL CALIBRATION

The execution of the model results in a series of new data layers illustrating the snapshots of simulated scenarios of the urban extent at the specified time frames. Such time frames include the final ending date of the model as well as dates at 10-year intervals from the starting date. With these results, users can compute the simulation accuracy of the model under certain configurations of the model's parameters and transition rules. Simulation accuracies include both the user's and producer's accuracies for each category of an urban state (non-urban, partly urban, and urban), as well as the overall accuracy of the model based on the modified error matrix approach presented in Section 5.2. In addition, the K_{hat} coefficient is also generated and delivered to the user for evaluation.

When computing the model's simulation accuracies, users have the choice to generate them only at the final iteration of the model, that is, the accuracies at the ending date of the model. However, users can also choose to compute simulation accuracies at each 10-year interval, for example, using the input data in 1986, 1996, and 2006 as reference data to evaluate the model's simulation accuracies for these respective years. These, at each 10-year interval, can provide additional information about the model's performance and can be used to calibrate the model temporally. This temporal calibration process will be exemplified in Chapter 6 when the model's results are presented for the Sydney Metropolitan Area.

Simulation accuracies of the model can be delivered to the users through the model calibration window (Figure 5.5); they can also be exported as a permanent document in Microsoft's Excel format for further analysis and evaluation.

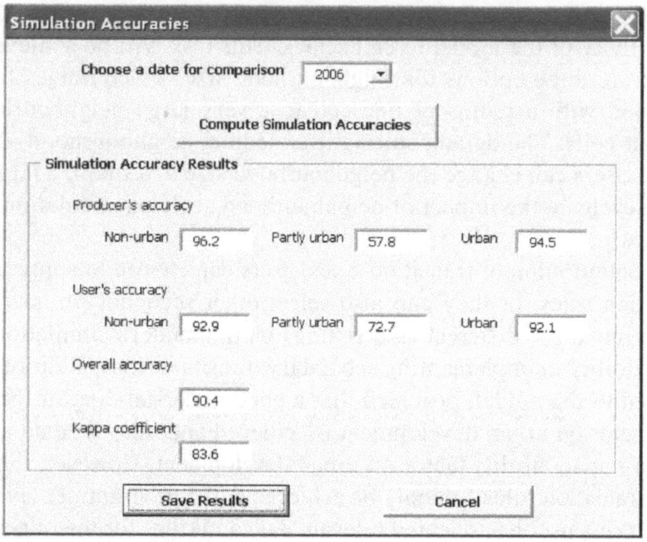

FIGURE 5.5 A snapshot showing the model's calibration window.

5.4 CONCLUSION

This chapter applies and implements the fuzzy constrained cellular automata model of urban development to Metropolitan Sydney, Australia. The five basic elements of cellular automata were specified with reference to the geographical data sets processed in Chapter 4. A number of primary and secondary transition rules reflecting the factors that affected the urban growth of Sydney from 1976 to 2006 were proposed; these factors were implemented in the model, based on the fuzzy constrained rule transition mechanisms. The various parameters associated with the transition rules as well as the scale of the neighbourhood will be modified during the model calibration process.

Implementation of the model was achieved within ESRI's ArcGIS program using VBA as the programming language. A simple GUI was designed to provide the flexibility for users to interact with the model by configuring not only the initial setting of the model but also the transition rules. The interface allows users to selectively test the impact of different transition rules on the model's behaviour, visualise its simulation output, and calibrate the model accordingly.

The model also provides the option for users to compute its simulation accuracies. These were achieved using a modified error matrix approach and the K_{hat} coefficient index. The simulation accuracies can be computed at different time periods, and the results can be returned to the user instantly for further calibration of the model. Results obtained from applying the model to Metropolitan Sydney in Australia are presented and discussed in Chapter 6.

6 Modelling the Urban Development of Sydney
Results and Discussion

This chapter presents the results of the fuzzy constrained cellular automata model in simulating the urban development of Sydney. By implementing different transition rules based on the principles of model calibration presented in Chapter 5, results generated by the model that match best with the actual urban development of the area are summarised in Section 6.1. Section 6.2 analyses the impact of individual urban development controls on the model's outputs, reflecting the strength of these controlling factors on Sydney's urban development from 1976 to 2006. The sensitivities of the model under different neighbourhood scales are analysed in Section 6.3, which results in further calibration and validation of the model in relation to different neighbourhood scales. Subsequently, the model is applied to generate prospective views of the future urban scenarios of Sydney under the proposed metropolitan planning strategies from 2006 to 2031. Finally, conclusions on the application of the cellular automata model for simulating Sydney's urban development are presented in the last section of this chapter.

6.1 A SUMMARY OF RESULTS FROM THE MODEL

6.1.1 THE SIMULATION AND CALIBRATION SEQUENCE OF THE MODEL

Based on the configuration comprising a cell size of 250 m and a circular neighbourhood with a radius of two cells, the model of urban development of Sydney was first implemented with the primary transition rules representing the contiguity effect; that is, the model concerns only the propensity of the cell itself for urban development and the support the cell may receive from its neighbourhood for such development. This resulted in an unconstrained model of urban development. Subsequently, secondary transition rules representing topographical constraints, transportation support, as well as urban planning controls were introduced into the model to calibrate it and evaluate its simulation accuracies using the past experience of urban development from 1976 to 2006 as references for comparison.

The secondary transition rules were introduced in the model based on the sequence of physical constraints, socio-economic factors, and institutional controls. Physical constraints such as water bodies, terrain, and other topographical conditions are "hard" constraints on urban development; these factors cannot be altered easily by human beings. In contrast, socio-economic factors and institutional controls are more "soft" in affecting the process of urban development, which can be changed with relative ease by human forces. Figure 6.1 shows a flowchart of the model's simulation and calibration sequence.

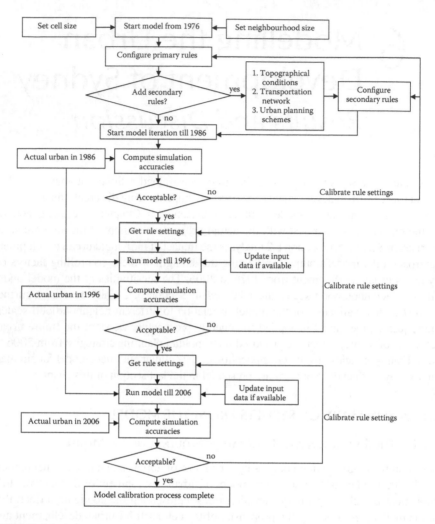

FIGURE 6.1 A flowchart showing the simulation and calibration sequence of the model.

6.1.2 OVERALL RESULTS UNDER ALL TRANSITION RULES

By varying the parameters of the transition rules within each iteration of the model and changing the balance between the rules, various simulation results were produced, and their accuracies against the actual urban development of Sydney at every 10-year interval were computed using the modified error matrix approach. Through vigorous calibrations of the model, the best results that the model has generated are illustrated in Figure 6.2e–g, which are compared with the actual scenarios of urban development from 1976 to 2006 (Figure 6.2a–d). The various simulation accuracies of these results are documented in Table 6.1. Figure 6.3 shows the various simulation accuracies of the model under all transition rules over time.

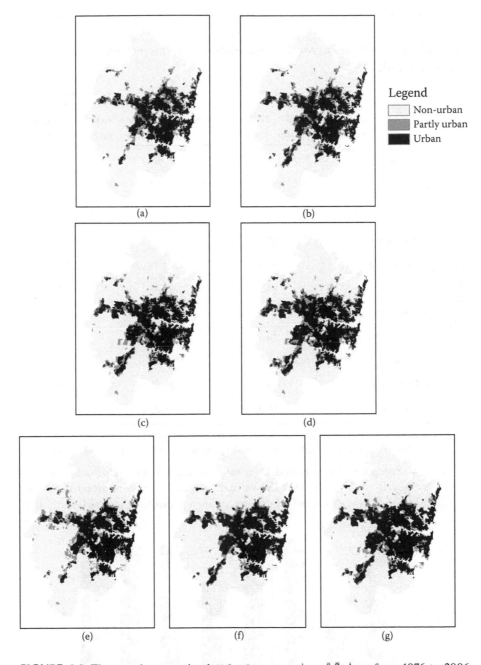

FIGURE 6.2 The actual versus simulated urban scenarios of Sydney from 1976 to 2006 under all transition rules: (a) actual scenario in 1976; (b) actual scenario in 1986; (c) actual scenario in 1996; (d) actual scenario in 2006; (e) simulated scenario in 1986; (f) simulated scenario in 1996; and (g) simulated scenario in 2006.

TABLE 6.1

The Urban Development Model's Simulation Accuracies under All Transition Rules

			1976	1986		1996		2006	
			Actual scenario	Actual scenario	Simulated scenario	Actual scenario	Simulated scenario	Actual scenario	Simulated scenario
Percentage of cells in each category (%)		Non-urban	71.4	67.8	69.0	64.9	68.0	63.2	64.6
		Partly urban	9.2	9.7	9.2	10.5	7.3	10.3	8.2
		Urban	19.4	22.5	21.8	24.6	24.7	26.5	27.2
Simulation Accuracies (%)	Producer's accuracy	Non-urban	–	93.4		97.3		96.2	
		Partly urban	–	53.7		51.9		57.8	
		Urban	–	88.2		92.4		94.5	
	User's accuracy	Non-urban	–	91.0		90.6		92.9	
		Partly urban	–	56.3		75.2		72.7	
		Urban	–	91.0		91.7		92.1	
	Overall accuracy		–	**86.8**		**89.5**		**90.4**	
	K_{hat} **Coefficient**		–	**76.3**		**81.5**		**83.6**	

Note: The areas excluded from urban development were not counted when calculating simulation accuracies. This is because the excluded areas were not participating in the model simulation process.

A visual comparison of the model's simulated results with actual urban scenarios of Sydney from 1976 to 2006 shows that, in general, results produced by the model matched well with the actual urban extent over this time period (Figure 6.2).

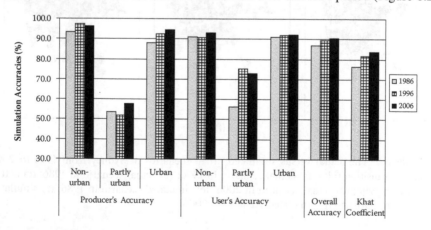

FIGURE 6.3 The model's simulation accuracies under all transition rules over time (1986–2006).

The spatial patterns of urban development generated by the model resemble the actual urban patterns to a large extent, especially for the year 2006. The various types of simulation accuracies produced by the model have been increasing steadily from 1976 to 2006. This is attributed to the calibration of the model over time. By the year 2006, the model had generated an overall accuracy of 90.4% and a K_{hat} coefficient of 83.6% (Table 6.1). This is significant given that the model only incorporated limited factors that contribute to the actual urban development of Sydney. Other factors such as the accessibility to nodes of employment and other services, and facilities such as schools, shops, public transports, etc., that may also have affected the process of Sydney's urban development had not been introduced into the model due to difficulties in data collection.

However, the producer's and user's accuracies of each category show that large discrepancies exist between the simulated results and the actual urban development in the partly urban category, resulting in lower accuracies from both the producer's and user's perspectives. For instance, the producer's and user's accuracies for the partly urban category in 2006 were only 57.8% and 72.7%, respectively. This means that 42.2% of the actual partly urban areas were omitted from being selected for development, and 27.3% of the simulated partly urban areas were committed to the category by the model incorrectly. In addition, for the same partly urban category, the producer's accuracy in 1996 was the lowest, whereas the user's accuracy was the highest among other years (Figure 6.3). This is also reflected in the model's results showing a lower percentage of 7.3% of partly urban areas and a higher percentage of 68% of non-urban areas in 1996, although the actual composition of partly urban and non-urban areas in the year were 10.5% and 64.9%, respectively. This is largely due to the fact that the partly urban areas only consist of approximately 10% of the total area of Sydney; therefore, even a small amount of mismatched cells between the actual urban extent and the simulated results would result in a high percentage of discrepancy between the two data sets, and hence, lower the producer's and user's accuracies in this category. On the other hand, due to the large composition of non-urban areas in the Metropolitan Sydney region, the relatively smaller portion of mismatched cells between the actual and simulated results may not have contributed significantly to omission or commission errors in this category as in the partly urban category.

For the fully urban category, because 19.4% of the Metropolitan Sydney region had already been fully urbanised at the start of the model in 1976, and these fully urban areas remain as urban during the whole simulation process, the actual composition of fully urban areas has only increased by 7.1% up to the year 2006. However, all urban cells including those that did not change states during the simulation process were counted when computing the simulation accuracies of the model for the fully urban category. This reduces the impact of the mismatched cells on both the producer's and user's accuracies for the fully urban category.

It should be noted that, when computing simulation accuracies throughout the whole process of model calibration, the areas excluded from urban development were not accounted for because those areas were not participating in the progression of the simulation process, making the simulation accuracies more realistic to reflect the actual performance of the model.

6.2 THE IMPACT OF INDIVIDUAL FACTORS ON SYDNEY'S URBAN DEVELOPMENT

Table 6.1 shows that under the primary and all secondary transition rules, the model had generated results with an overall accuracy of 90.4% and a K_{hat} coefficient of 83.6% by the year 2006. However, the impact of individual factors on the urban development of Sydney over time varies, which needs to be evaluated by using the model. Based on the calibration principles of the model, the secondary transition rules were introduced into it progressively in the sequence of physical constraints, socio-economic support, and institutional controls. Hence, it is possible to evaluate the impact of individual factors on the model's performance and outcomes in such sequence. Table 6.2 presents the progression of the model's simulation accuracies under different transition rules in 2006. Figure 6.4 is a graphical representation of the results presented in Table 6.2.

Table 6.2 and Figure 6.4 show that, by introducing each new transition rule into the model, simulation accuracies increased. Even though the improvement in the overall simulation accuracy of the model was only from 86.3% when the model was configured with primary transition rules to 90.4%, when all primary and secondary transition rules were implemented in the model, the improvements in the individual producer's and user's accuracies were significant, especially for the partly urban category. This also resulted in a significant increase of the K_{hat} coefficient from 76.1 to 83.6%. The steady increase in the model's simulation accuracies was also reflected in the overall accuracy and the K_{hat} coefficient over time (Figure 6.5).

TABLE 6.2

The Model's Simulation Accuracies under Different Transition Rules in 2006

				Actual scenario in 2006	Primary rules (PRs) only	PRs plus topographical constraints	PRs plus topographical constraints and transportation support	PRs plus topographical constraints, transportation support, and urban planning
Percentage of cells in each category (%)			Non-urban	63.2	67.6	67.4	63.1	64.6
			Partly urban	10.3	7.6	6.7	9.5	8.2
			Urban	26.5	24.8	26.0	27.4	27.2
Simulation accuracies (%)	Producer's accuracy		Non-urban	–	95.4	94.8	92.6	96.2
			Partly urban	–	46.8	46.8	57.5	57.8
			Urban	–	88.0	90.8	94.2	94.5
	User's accuracy		Non-urban	–	85.7	88.3	92.6	92.9
			Partly urban	–	63.6	60.8	62.5	72.7
			Urban	–	94.1	92.8	91.4	92.1
	Overall accuracy			–	86.3	87.0	88.4	90.4
	K_{hat} coefficient			–	76.1	77.5	80.4	83.6

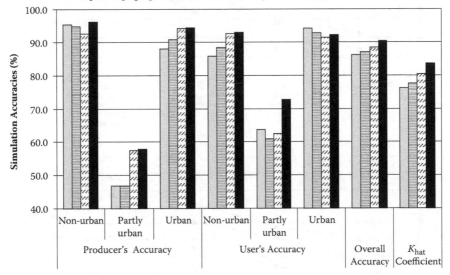

FIGURE 6.4 The model's simulation accuracies in 2006 under different transition rules.

Based on the progression of the secondary transition rules being introduced into the model, the following sections evaluate the impact of individual factors on the urban development of Sydney from 1976 to 2006.

6.2.1 Unconstrained Urban Growth

The unconstrained urban growth of Sydney was generated by applying the primary transition rules only to the model; that is, this version of the model only deals with the propensity of a cell for development and the support for such development that the cell may receive from its neighbourhood. No secondary transition rule was introduced into the model to constrain or accelerate the natural process of urban development.

Results generated by the model with the configuration for unconstrained urban growth show that many partly developed cells had been further developed over the simulation period from 1976 to 2006, some of them becoming fully urban areas. However, developments from non-urban to partly urban or fully urban cells were extremely slow. As a result, most urban developments that actually occurred in the west and south-west parts of Sydney over the simulation period were not represented in the model's results. Figure 6.6 illustrates the discrepancy of areas between the actual urban extent and the simulated results from the model in 2006. This figure shows that large patches of land that had actually been developed in the north-west and south-west parts of Sydney were omitted by the model, with some smaller patches of land scattered over the existing urban areas of Sydney committed incorrectly as developed, either as urban or partly urban areas.

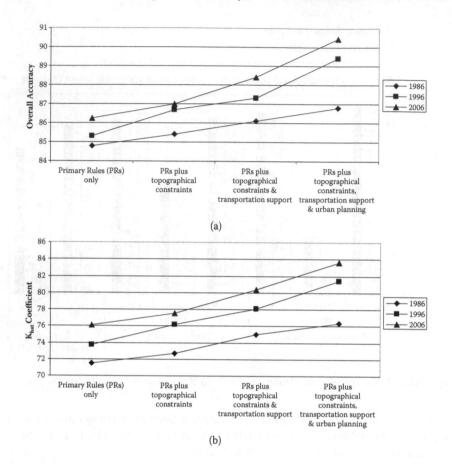

FIGURE 6.5 The overall simulation accuracy (a) and the K_{hat} coefficient (b) generated by the model under different transition rules over time.

The spatial accuracies of the simulation results of 2006 show that, although the overall agreement between the simulated and actual urban growth of Sydney reached 86.3%, significant discrepancies between the two data sets exist in each category of the cell states (Table 6.2). For instance, the simulated results show that 67.6% of the total areas were in the non-urban state in 2006, whereas 7.6% and 24.8% were in partly urban and urban states, respectively. However, the actual urban composition was 63.2% for non-urban areas, 10.3% for partly urban areas, and 26.5% for urban areas.

In addition, the producer's accuracy for the partly urban category was only 46.8% and the user's accuracy was also as low as 63.6%. This shows that over half, that is, 53.2% of the partly urban areas that had been developed on the ground had been omitted from this category by the model. On the other hand, 36.4% of the partly urban areas identified by the model were committed incorrectly into this category. For the urban category, an omission error of 12% and a commission error of about 6% were also observed.

FIGURE 6.6 Spatial discrepancies between the actual urban extent and the simulated results from the model in 2006 under unconstrained urban growth conditions. The figure shows that large patches of developed or partly developed land were omitted by the model.

Comparing the omission and commission errors of each category individually, it was the omission errors that were dominant in both the urban and partly urban categories, and the commission error that was dominant in the non-urban category, indicating that the model did not generate sufficient development to match the actual urban development of Sydney. Therefore, some cells were omitted from the urban or partly urban category and committed to the non-urban category incorrectly. As such, the K_{hat} coefficient was only 76.1%. Even by changing the initial setting of the parameters and the balance between the transition rules as well as calibrating the model over time, the model was not able to launch sufficient development from the non-urban cells. The discrepancy between the simulation results and the actual urban development indicated that there were other conditions that had promoted continuous urban growth or generated new development.

6.2.2 Topographically Constrained Development

This version of the model introduced a set of secondary transition rules reflecting the topographical constraints on the urban development of Sydney. The model first introduced land slope as a constraining factor. Subsequently, studies of the Sydney region show that urban development in the southern part of Sydney near Menai was largely driven by the relatively high terrain, beautiful natural environment, and close proximity to the east coast. Geographically, this area is situated in the transition between the Woronora Plateau and the Cumberland Plain. Most of the area has a terrain about 100 m above sea level. Thus, residents can enjoy the beautiful views of the sea and the coast. To represent these conditions, transition rules were introduced in the model, representing the attractiveness of the terrain and coastal proximity to urban development.

By calibrating the model's parameters with the urban area data sets of Sydney from 1976 to 2006, a different scenario of the urban development of Sydney was generated, resulting in slightly better accuracies of the model in all categories (see Table 6.2). However, results from this version of the model still show high omission and commission errors, especially in the partly urban category. Even though the model was capable of controlling the urban development in areas having a steep terrain and initiating new development to some extent in areas having advantageous terrain and coastal proximity attractions, the model was not able to generate enough development to match the actual development that occurred on the ground. Hence, it was necessary to introduce new rules for the transition of the cells from the non-urban to partly urban and urban states.

6.2.3 Transportation-Supported Development

Next, a number of secondary transition rules were added to the model to represent the support of the transportation network on urban development in Metropolitan Sydney from 1976 to 2006. These transition rules have resulted in different patterns of urban development scenarios in Sydney. Figure 6.7 illustrates three snapshots of the results generated by the model in 1986, 1996, and 2006, respectively.

Comparing the model's results with the actual urban development of Sydney over the same time period (see Figure 6.2b–d), a good similarity between the simulated results

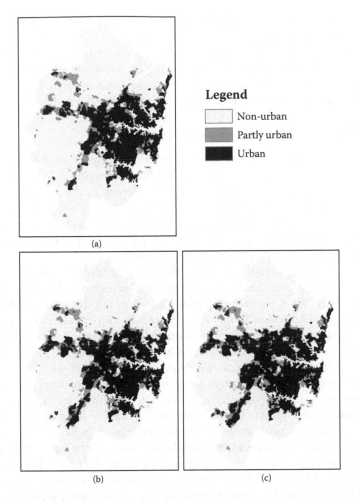

FIGURE 6.7 Snapshots of the urban scenarios generated by the model with topographical constraints and transportation support in (a) 1986, (b) 1996, and (c) 2006.

and the actual urban extent can be observed when rules representing both topographical conditions and transportation network support were implemented in the model. A poorer similarity between the model's results and the actual urban extent exists during the early calibration stage from 1976 to 1986; however, this improved when the model was calibrated toward the later stages. Geographically, the model generated most of the urban developments that occurred in the west and south-west parts of Sydney.

The simulation accuracies of the results generated by this version of the model show that, with both topographical constraints and transportation support, the model generated an overall accuracy of 88.4% and a K_{hat} coefficient of 80.4% (see Table 6.2). This shows a better match of the model's results with the actual urban development on the ground. In particular, the individual accuracy of each category improved by various degrees, with a significant increase in the partly urban category in both the producer's and user's accuracies. As a result, the omission error of the partly urban cells was reduced from 53.2% in the previous version of the model to 42.5% in the current version. The omission error of the urban category also decreased from 9.2 to 5.8%. On the other hand, commission errors for the partly urban and urban categories have also decreased to 37.5% and 8.6%, respectively. These improvements indicate that the model generated more development that matches the actual urban development in Sydney.

6.2.4 Urban Planning Policies and Schemes

The last version of the model was produced by adding another set of secondary transition rules representing the impact of various urban planning policies and schemes on the urban development of Sydney over the lifespan of the respective planning schemes. As previously discussed, urban planning schemes were introduced in the model in two ways. One was by initiating more development in areas proposed for development, and the other by controlling the overall speed of urban development based on the nature of the respective planning schemes.

By adding this set of transition rules to the model, its overall simulation accuracy increased to 90.4%, and the K_{hat} coefficient also increased to 83.6% in 2006 (see Table 6.2). This shows a close resemblance between the simulated results and the actual urban development in Sydney (see Figure 6.2). In particular, although the producer's accuracies in each category have only slightly changed, the user's accuracy in the partly urban category has increased by over 10%, resulting in a low commission error of only 27.3% in this category (see Table 6.2 (d) and (g)).

It should be noted that even though areas proposed by each of the planning schemes were introduced into the model as accelerating factors to initiate new or speed up existing development, the impact of these factors was weak. For instance, the 1968 urban planning scheme was implemented in the model from 1976 to 1987. Based on this planning scheme, areas proposed in principle to be urban in the future were identified as having a higher probability to be developed. However, the calibration of the model shows that this "higher probability" can only be represented by an increase of around 40 to 50% in the value of the model's planning control parameter, which only became effective when this factor worked in conjunction with other conditions such as self-propensity and neighbourhood support, or with other secondary transition conditions.

For the 1988 urban planning scheme, the effective lifespan of the plan was from 1988 to 1994. However, no significant development had occurred in the areas under consideration during the lifespan of the plan. Therefore, apart from assessing all other conditions such as self-propensity for development and neighbourhood support, as well as topographical conditions and transportation support, no additional probability was accredited to this planning scheme for new or continuous development. Instead, a global parameter was assigned to increase the rule-firing threshold by 20%, that is, from 0.5 to 0.6, reflecting the preference of the plan in limiting urban sprawl and promoting urban consolidation and redevelopment.

For the 1995 urban planning scheme, the effective lifespan of the plan on Sydney's development started from 1995 until the new metropolitan strategy was proposed in 2005. Based on the plan, areas proposed for development in the urban development programmes were assigned a higher probability for development than other areas;

this probability was configured at a percentage ranging from 40 to 60%, subject to other conditions. In addition, a higher rule-firing threshold of 0.6 was also applied globally to reflect the overall control on urban expansion and the promotion of urban consolidation and redevelopment.

Should any of the parameters related to urban planning schemes and policies be changed even slightly in the model, major discrepancies between the simulated results and the actual urban development on the ground would follow, which may result in low simulation accuracies of the model. This is because not all areas planned for urban development have actually been developed over the proposed time periods. Urban development occurred only in areas having a certain degree of accessibility and services, such as a relatively flat terrain and support for development from the transportation infrastructure. Hence, to ensure that development will actually occur as proposed in the planning schemes, it is important to improve the level of accessibility to such areas and the spatial connection communication between the proposed and other developed areas.

6.2.5 OTHER TRANSITION RULES

So far, the model has tested the impact of a number of secondary transition factors on the process of urban development in Sydney from 1976 to 2006. Each of these factors, including the physical landscape constraints, transportation network, as well as urban planning programmes, was identified as a major contributor to the urban development of Sydney. Although it was difficult to separate the contribution of each individual factor to this development, it was clear that transportation network played the most significant role among all other secondary transition rules. In addition, urban planning as represented in this model also contributed to Sydney's development. Only some of the proposed urban development areas that were supported by a certain degree of transportation infrastructure and services have been developed. The urban planning programs have also contributed to Sydney's development by releasing the constraint of topographical conditions and reinforcing the impact of transportation network on this development.

However, not all factors affecting Sydney's urban development have been considered in the model. For instance, although the transportation network has been implemented in the model to represent accessibility, factors such as journey to work and access to other services and facilities including schools, shops, sewerage, and drainage systems were not modelled. This is partly due to the difficulty in collecting the spatial and non-spatial data to quantify the impact of such factors on the urban development of Sydney. In this regard, the model needs to be fine-tuned when more spatial data become available.

Nonetheless, the model has demonstrated a flexible way in which different rules can be implemented in the cellular automata model to control the simulation of urban development. Other factors, such as urban infrastructure (drainage and sewerage systems), incomes, and community services, can also contribute to the urban development of Sydney. Some of these factors could be important to local urban development. Nevertheless, the model has produced realistic results in illustrating Sydney's urban development.

With the flexibility of rule implementation within the model, more rules can be added to fine-tune it, provided that a good understanding of the rules is maintained and good data are collected. The flexibility of the model enables it to function not only as an analytical tool to understand the factors controlling the process of urban development but also as a planning tool to experiment with various planning proposals and answer the "what if" questions in the planning practice.

6.3 THE IMPACT OF NEIGHBOURHOOD SCALE ON THE MODEL'S RESULTS

Previous sections presented results of the model under various transition rules. In this modelling process, the cellular automata model of urban growth was configured with a circular neighbourhood having a radius of two cells. This neighbourhood size was not configured by intuition but through a number of experiments and comparisons with other neighbourhood sizes. To demonstrate the impact of neighbourhood size on the model's results, this section presents results from three different neighbourhood sizes: the first with a radius of one cell representing a small neighbourhood, the second with a radius of three cells representing a large neighbourhood, and the third with a radius of four cells representing a very large neighbourhood (Figure 5.1a,c,d in Chapter 5). The initial neighbourhood setting of a radius of two cells (Chapter 5, Figure 5.1b) is therefore termed a medium-size neighbourhood.

6.3.1 RESULTS FROM THE MODEL UNDER DIFFERENT NEIGHBOURHOOD SCALES

Using the small, large, and very large neighbourhood sizes, the model was configured with both primary and secondary transition rules in the same way as discussed in Section 6.2.4. The results generated by the model under these three different neighbourhood sizes are displayed in Figure 6.8, and their simulation accuracies in comparison with the accuracies from the medium-size neighbourhood are summarised in Table 6.3.

Comparing the results generated by the model under small, large, and very large neighbourhood sizes (Figure 6.8) with results under the medium-size neighbourhood and the actual urban development of Sydney (Figure 6.2), the model with the small- and medium-size neighbourhoods generated similar patterns of urban scenarios; their simulation accuracies were very close (Table 6.3). However, with a large or a very large neighbourhood configuration, the model generated significantly more development than the actual urban development on the ground.

Statistically, by calibrating the model temporally, the model was able to generate an overall accuracy of 90.1% and a K_{hat} coefficient of 83.2% under a small-neighbourhood configuration. These values were very close to the results generated by the model under the medium-size neighbourhood. This is because the small-neighbourhood scale used eight cells surrounding the central cell in question, whereas the medium-size neighbourhood only added an additional three cells in each direction (East, North, West, and South). Both neighbourhood sizes measured the impact of the immediate neighbourhood on the development of the central cell in question. However, with the small-size neighbourhood, the user's accuracies for both the partly urban and urban categories were lower when compared to the medium-size neighbourhood

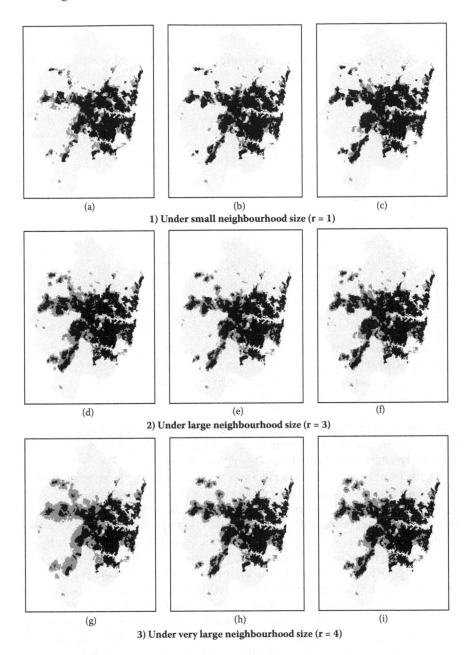

(a)   (b)   (c)

1) Under small neighbourhood size (r = 1)

(d)   (e)   (f)

2) Under large neighbourhood size (r = 3)

(g)   (h)   (i)

3) Under very large neighbourhood size (r = 4)

FIGURE 6.8 Simulated urban scenarios of Sydney under different neighbourhood scales. (1) Under small neighbourhood size (r =1): (a) 1986, (b) 1996, (c) 2006; (2) under large neighbourhood size (r = 3): (d) 1986, (e) 1996, (f) 2006; under very large neighbourhood size (r = 4): (g) 1986, (h) 1996, and (i) 2006. (*Note*: The legend is the same as in Figure 6.7).

TABLE 6.3

The Model's Simulation Accuracies under Various Neighbourhood Scales in 2006

			Actual scenario	Small neighbour- hood	Medium neighbour- hood	Large neighbour- hood	Very large neighbour- hood
Percentage of cells at each category (%)		Non-urban	63.2	63.6	**64.6**	60.4	58.6
		Partly urban	10.3	8.7	**8.2**	12.6	14.4
		Urban	26.5	27.7	**27.2**	27.0	27.0
		Total	100	100	**100**	100	100
Simulation accuracy assessment (%)	**Producer's accuracy**	Non-urban	–	95.1	**96.2**	89.0	84.9
		Partly urban	–	59.5	**57.8**	70.8	72.5
		Urban	–	94.6	**94.5**	94.1	94.1
	User's accuracy	Non-urban	–	94.0	**92.9**	95.6	96.0
		Partly urban	–	70.3	**72.7**	58.3	52.0
		Urban	–	90.8	**92.1**	92.3	92.3
	Overall accuracy		–	90.1	**90.4**	88.3	86.4
	K_{hat} **coefficient**		–	83.2	**83.6**	80.7	77.9

configuration, even though the producer's accuracy for the partly urban category was higher. This shows that there were more commission errors than omission errors generated by the model under the small-neighbourhood scale.

On the other hand, when the model was configured with a large-neighbourhood scale of a three-cell radius, the overall accuracy of the model was reduced to 88.3%, and the K_{hat} coefficient was reduced to 80.7% by the year 2006. These accuracy values were further decreased to 86.4% and 77.9%, respectively, when a very-large-neighbourhood scale was applied to the model. In particular, with the large- and very-large-neighbourhood configurations, the producer's accuracies at the partly urban category increased significantly to over 70%, whereas for the non-urban category this accuracy was reduced to less than 90%. On the contrary, the user's accuracies for the partly urban category decreased significantly to less than 60%, whereas for the non-urban category the user's accuracies under the large- and very-large-neighbourhood scales were higher. The results show that, under the large- and very-large-neighbourhood configurations, the model generated more development than the actual development on the ground. Hence, the model overconsidered the impact of a large or very large neighbourhood on local urban development.

Owing to the discrepancies of the individual accuracies in each category from both the producer's and user's perspectives, the overall simulation accuracies of the model under the large- and very-large-neighbourhood scales were only 88.3% and 86.4%, respectively, in 2006, and the K_{hat} coefficients were as low as 80.7% and 77.9%, respectively (Table 6.3).

6.3.2 SIMULATION ACCURACIES OF THE MODEL OVER TIME

Similar results were observed when examining the model's performance under different neighbourhood scales over time (Table 6.4 and Figure 6.9). The overall simulation accuracies and the K_{hat} coefficient values for the small- and medium-neighbourhood scales were close, which also demonstrated consistent improvement over time. However, under the configurations of the large- and very-large-neighbourhood scales, lower simulation accuracies were observed during the whole calibration process of the model, with the lowest accuracy values observed around the year 1986 (when the model was configured with a large-neighbourhood scale) and around the year 1996 (when the model was configured with a very-large-neighbourhood scale). The reason behind this temporal variation needs further investigation.

TABLE 6.4

The Model's Simulation Accuracies under Different Neighbourhood Scales over Time

		Producer's accuracy			User's accuracy				
		Non-urban	Partly urban	Urban	Non-urban	Partly urban	Urban	Overall accuracy	K_{hat} coefficient
Small neighbourhood	1986	92.0	58.7	88.6	92.5	54.7	90.4	86.8	76.6
	1996	96.3	54.6	92.6	91.9	72.7	90.3	89.4	81.5
	2006	95.1	59.5	94.6	94.0	70.3	90.8	90.1	83.2
Medium neighbourhood	1986	93.4	53.7	88.2	91.0	56.3	91.0	86.8	76.3
	1996	97.3	51.9	92.4	90.6	75.2	91.7	89.5	81.5
	2006	96.2	57.8	94.5	92.9	72.7	92.1	90.4	83.6
Large neighbourhood	1986	77.1	56.5	99.7	97.8	39.2	81.2	81.1	68.9
	1996	90.5	66.2	92.3	94.1	58.6	91.6	87.7	79.3
	2006	89.0	70.8	94.1	95.6	58.3	92.3	88.3	80.7
Very large neighbourhood	1986	87.1	68.7	92.2	94.6	53.0	91.5	86.2	77.1
	1996	78.3	74.9	88.2	96.6	38.3	91.7	80.8	68.7
	2006	84.9	72.5	94.1	96	52	92.3	86.4	77.9

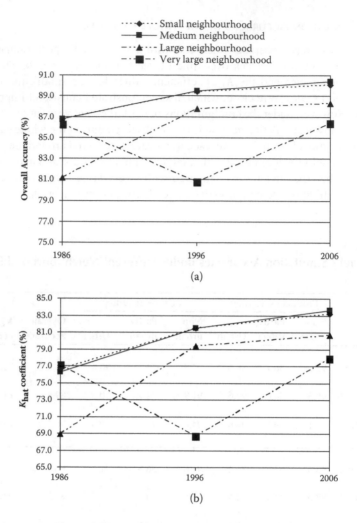

FIGURE 6.9 The overall simulation accuracy (a) and the K_{hat} coefficient (b) of the urban development model under different neighbourhood scales over time. The model generated higher simulation accuracies under small- and medium-neighbourhood scales; these accuracies also improved consistently over time. However, significant differences between the simulated results under large or very-large neighbourhoods and the actual urban development exist, especially around the years 1986 (when the model was configured with a large-neighbourhood scale) and 1996 (when the model was configured with a very-large-neighbourhood scale).

As urban development is controlled by locally defined transition rules in a cellular automata model, the driving force for development a cell can receive from its neighbourhood is limited to a certain distance. In the case of the urban development of Sydney, this distance is around two cells from the x- and y-directions of the central cell in question, or a 500-m radius surrounding each locality. This neighbourhood size corresponds to a local block or neighbourhood community. Increasing the size of the neighbourhood resulted in extended areas that drive the development of a cell.

This driving force may not exist in actual urban development. Therefore, the model generated overdevelopment. On the other hand, decreasing the neighbourhood size may reduce the size of areas that can drive the development of the cell. Hence, the model may generate scenarios of underdevelopment.

6.3.3 NEIGHBOURHOOD SCALE AND MODEL CALIBRATION

The introduction of cellular automata modelling to urban studies is an applied science, and the neighbourhood scale plays an important role in the configuration of the model and its performance. Although previous researches applied both large and small neighbourhood sizes in modelling urban growth using the cellular automata approach (Li and Yeh 2000; Clarke and Gaydos 1998; Wu 1998a,b,c, 1996; Clarke, Hoppen, and Gaydos 1997; White and Engelen 1994, 1993), no particular validation of the neighbourhood scale has been explored.

Through the comparative study of various neighbourhood sizes on simulating the urban development of Sydney, a moderate-neighbourhood scale was selected to represent a local community that affects local urban development. This neighbourhood scale may not be applicable to other areas. However, as it is one of the fundamental elements in cellular automata-based urban modelling, it is important to understand the size of the neighbourhood by which a cell can be affected and examine the variation in sizes of such neighbourhood on the model's behaviour and outcomes. By testing and calibrating the model under different neighbourhood scales, a suitable neighbourhood size may be identified for a cellular automata model to simulate the process of urban development.

6.4 PERSPECTIVE VIEWS ON SYDNEY'S DEVELOPMENT TO THE YEAR 2031

Sydney has been constantly developing over time. The preceding sections constructed a cellular automata model of urban development and applied the model to understand the factors controlling the urban development of Sydney from 1976 to 2006. With the setting of a medium-neighbourhood scale and the transition rules representing natural urban growth, topographical constraints, transportation support, as well as urban planning controls, the model has generated results that resemble the actual urban development of Sydney. An understanding of how the city has been growing will assist in projecting the future directions of this growth, provided the factors that may emerge to affect such growth in the future are well considered in the model. The following sections apply the fuzzy constrained cellular automata model to generate perspective views on the future extent and composition of the urban system in Sydney. Here, the composition of the urban system refers to the structure of the non-urban, partly urban, and urban components of the system.

6.4.1 FACTORS AFFECTING SYDNEY'S FUTURE DEVELOPMENT

In addition to the factors previously discussed that have been driving the urban growth of Sydney since the 1970s, including the self-propensity and neighbourhood support, topographical constraints, transportation support, as well as urban planning controls, a number of new drivers have emerged since 2006 that may affect the urban development in Sydney in the years to come. The most predominant factors are identified in the following sections, which will be implemented in the model to generate perspective views of urban development in Sydney from 2006 to 2031, the same time frame as the 2005 Sydney Metropolitan Strategic Plan.

6.4.1.1 Improvement in Transportation Infrastructure

A significant amount of transportation infrastructure has been constructed or is currently under construction in Metropolitan Sydney since 2006. This improvement in transportation infrastructure will have a significant impact on the urban development of the city in the future. Hence, data need to be collected to update the previously processed transportation index data when the model is applied to generate perspective views on future urban development in Sydney.

One of the most significant improvements in the road infrastructure is the completion of the Westlink M7 Motorway in 2006. The construction of the motorway is to serve the rapid growth in western Sydney. This motorway starts at the Hume Highway/M5 South Western Motorway interchange at Prestons, passes through the western Sydney suburbs of Liverpool, Fairfield, Blackrown, and Baulkham Hills, and ends at Baulkham Hills where it joins the M2 Hills Motorway (see Chapter 4, Figure 4.8). The total length of the M7 Motorway is 40 km, which forms a part of the Sydney Orbital Motorway network. Through a number of interchanges to link the intercity highways to the Orbital network, the M7 Motorway provides an alternative to move traffic away from other busy routes, which can reduce transit time across the western suburbs by one hour or more.

Another important improvement is the construction of the railway line that connects Epping station on the Northern line to Chatswood station on the North Shore line. The construction started in 2002 and is scheduled to be completed by the end of 2008. With major redevelopment at the two junction stations at Epping and Chatswood, there will also be three new stations at North Ryde, Macquarie Park, and Macquarie University, respectively.

In addition, a further extension of the Epping-to-Chatswood railway line to the north-west part of the metropolitan area will begin in 2010. This line is planned to open in two stages; the first stage is from Epping to Hills Centre, which is due to be completed by 2015, and the second stage is from Hills Centre to Rouse Hill, which is to be completed by 2017. The total length is 37 km, with 17 new railway stations servicing the suburbs of St. James, Martin Place, Wynyard, Pyrmont, Top Ryde, Epping, Castle Hill, Hills Centre, and Rouse Hill.

Yet another commuter railway line that is planned is the South West Rail Link, connecting Glenfield and Leppington. This railway line is designed to cater to the growth in the South West Growth Centre proposed in the 2005 Metropolitan Strategic Plan. The total length is 12 km, with two new stations at Leppington and Edmondson Park, respectively. The project also includes a train stabling facility to the west of the new Leppington station. According to the plan, the construction of the project will begin in 2009 and is expected to be completed in 2012.

Data reflecting this transportation infrastructure development are collected and processed to update the transportation data sets of the model. The time lines in the construction of the transportation network are also considered; hence, a temporal database reflecting the constant change of transportation infrastructures is maintained and used by the model.

6.4.1.2 The Impact of the 2005 Metropolitan Strategic Plan

Another important factor that will affect the future urban development of Sydney is the 2005 Metropolitan Strategic Plan (New South Wales Department of Planning 2005). As discussed in Chapter 4, Section 4.1.2.5, the Metropolitan Strategic Plan has identified a global economic corridor, a hierarchy of cities and urban centres, as well as two new growth centres. Areas around transportation routes that connect cities and urban centres are also identified as development corridors. To reflect the impact of the strategic plan on the future urban development of Sydney, a number of new data sets including a point data set for cities and urban centres and a polygon data set illustrating the North-West and South-West Growth Centres were collected and processed in ArcGIS. The development corridors were identified by generating buffer zones around the Sydney Orbital Motorway Network and along the major railway lines (see Chapter 4, Figure 4.8).

Similar to the way the urban planning control factor was introduced into the cellular automata model as secondary transition rules during the model calibration process, the possible impact of the Sydney Metropolitan Strategic Plan was introduced into the model in two ways: the first, to act as an accelerating factor to promote urban development along the development corridors, cities, and urban centres, which is achieved by reducing the rule-firing threshold in the model; and the second, to initiate new development within the growth centres, especially those areas with good accessibility to transportation infrastructure and services.

Based on the assumption that the 2005 Metropolitan Strategic Plan may play a weak or strong role in guiding the urban development of Sydney, the model has been applied to generate two perspective views of the year 2031, one under weak planning controls and the other under strong planning controls.

6.4.2 Perspective Views of Urban Development under Different Planning Control Factors

According to the 2005 Metropolitan Strategic Plan, a hierarchy of regional cities, specialised centres, major centres, and planned and potential major centres were assigned different weights representing the impact of these cities or centres on the urban development of a region. Under the assumptions that urban planning will play a weak or strong role in guiding the urban development of Sydney to the year 2031, two different sets of weights were assigned to the centres to represent the strength of the centres affecting urban development, which are listed in Table 6.5.

Centres having large weights will have a larger coverage of areas to impact upon and also more strength to drive the urban development of the region, whereas centres having small weights will have a smaller coverage of areas to impact upon with weaker strengths. However, with the development of the centres themselves,

TABLE 6.5

Weights Assigned to the Hierarchy of Urban Centres under the Weak and Strong Planning Control Conditions

		Weight	
Type	Centre names	Under weak planning controls	Under strong planning controls
Global Sydney	Sydney City, North Sydney	5	10
Regional city	Parramatta, Liverpool, Penrith	4	8
Major centre	Bankstown, Blacktown, Bondi Junction, Brookvale-Dee Why, Burwood, Campbelltown, Castle Hill, Chatswood, Hornsby, Hurstville, Kogarah	3	3
Specialised centre	Macquarie Park, St Leonards, Olympic Park-Rhodes, Port Botany, Sydney Airport, Randwick Education and Health, Westmead, Bankstown Airport-Milperra, Norwest	2	4
Planned major centre	Rouse Hill, Leppington, Green Square	2	4
Potential major centre	Sutherland, Cabramatta, Mt Druitt, Fairfield, Prairiewood	1	2

a planned major centre may gain more weight to impact the urban development of neighbouring areas. This is the same for the potential major centres.

The overall impact of the urban planning factor on urban development is reflected in the model through the setting of the rule-firing threshold (see Chapter 3, Section 3.3.3.2 for more discussion on the rule-firing threshold). Under the assumption of the weak planning control conditions, the rule-firing threshold for urban development is set to 0.5, the value that was calibrated in the previous sections to generate views that matched well with the actual urban development on the ground. For development under the strong planning control conditions, the rule-firing threshold was set 20% lower. This implies that under the strong planning control factors, areas identified by the plan, which also satisfy other development conditions, will have a 20% higher chance of being developed than those identified under the weak planning control conditions.

With the aforementioned assumptions and the updated transportation data set, the model was implemented using the actual urban extent of Sydney in 2006 (as a starting date) to generate perspective views of Sydney up to the year 2031. Figure 6.10 illustrates snapshots of the urban scenarios of Sydney in 2021 and 2031 under the two different assumptions of urban planning control factors. Figure 6.11 illustrates areas developed from 2006 to 2031 under these two different assumptions.

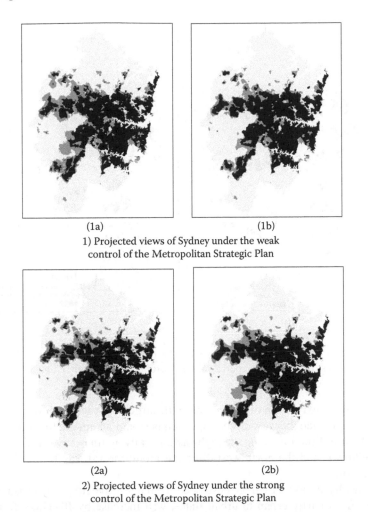

(1a)   (1b)

1) Projected views of Sydney under the weak
control of the Metropolitan Strategic Plan

(2a)   (2b)

2) Projected views of Sydney under the strong
control of the Metropolitan Strategic Plan

FIGURE 6.10 Perspective views on Sydney's development. (1) Projected views under weak control: (1a) 2021 and (1b) 2031. (2) Projected views under the strong control: (2a) 2021 and (2b) 2031.

Comparing the simulated scenarios of urban development in 2021 and 2031 with the actual urban extent in 2006, it is clear that urban development will continue in two ways: one is through the in-filling of areas within the currently existing urban or partly urban areas, and another is through expanding toward the west and south-west of the metropolitan areas (Figure 6.11). Through the in-filling process, some of the vacant land within the existing urban areas will be developed into partly urban or even urban areas, and some areas that are already developed to a certain extent will continue to be further developed into a fully urban state. Results from the model show that large patches of land that will be developed can be seen around the Sydney Airport area in the east, the Sydney's Olympic site at the Homebush Bay area, and in the south part of Sydney near Menai.

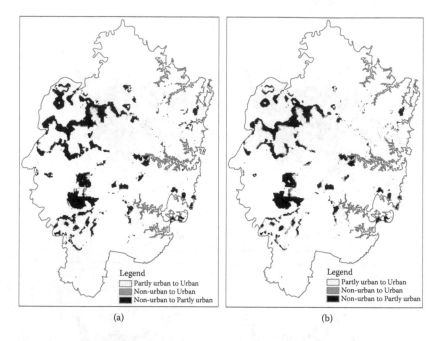

FIGURE 6.11 A comparison of areas that will be developed from 2006 to 2031 under the (a) weak and (b) strong planning control of the Metropolitan Strategic Plan.

The outward expansion of urban areas will largely be seen in three major directions. These include the west direction, which is from Parramatta to around Penrith; the south-west direction along the highway and railway lines toward Liverpool and Campbelltown; and the north-west direction along the railway line to Riverstone and Windsor.

Statistically, areas that will be developed from non-urban to partly urban or urban states, or from partly urban to urban states, will increase by 10–11.6% from 2006 to 2031, depending on the strength of the Metropolitan Strategic Plan in controlling their development. Subsequently, non-urban areas will decrease by the same percentage. Table 6.6 presents results of the projected urban area compositions from 2011 to 2031 under the two different assumptions of planning control conditions mentioned earlier.

However, by comparing the two sets of results under the different strengths of urban planning controls, the results under the weak planning controls generated more development in the partly urban category, although areas in the fully urban category will be 2.7% less compared to the urban areas under the strong planning control factors by the year 2031. On the other hand, the percentage of non-urban areas under weak planning control factors is 1.6% less than that under the strong planning control factors, and areas in the partly urban category will be 4.3% higher. This indicates that under the weak planning control factors, more non-urban areas will be chosen for development. However, the extent of development will be less intensive; hence, more areas will be in the partly urban state by the year 2031.

TABLE 6.6

Projected Urban Area Compositions from 2011 to 2031 under the Weak and Strong Planning Control Conditions

	Percentage of each category under the weak planning controls (%)					
	2006 (actual)	2011	2016	2021	2026	2031
Non-urban	63.1	60.6	58.3	56.9	54.6	53.1
Partly urban	10.3	10.8	13.0	14.1	15.6	16.7
Urban	26.5	28.5	28.7	29.0	29.9	30.2
	Percentage of each category under the strong planning controls (%)					
	2006 (actual)	2011	2016	2021	2026	2031
Non-urban	63.1	60.6	59.0	58.0	56.0	54.7
Partly urban	10.3	10.9	11.1	11.0	11.8	12.4
Urban	26.5	28.5	29.9	31.0	32.2	32.9

Compared with the results under the weak planning control factors, those under the strong planning control factors show that urban development will be more intensive, especially after the first 5 years from 2006 to 2011; the effect of the Metropolitan Strategic Plan will start to show from 2011 onward. By the year 2031, the composition of non-urban areas under strong control factors will be 1.6% higher than that under weak planning control factors, indicating that about 60 km² of land will be saved from urban development. However, the reduced amount of land for development does not mean less development but that it is compensated by a lower percentage of the partly urban areas and a higher percentage of the fully urban areas. Therefore, more rural land will be saved under the strong planning control conditions, and more partly developed urban areas will undergo further development to reach the fully urban state.

Overall, urban development in Sydney will continue through both in-filling in existing urban areas and outward expansion toward the west, south-west, and north-west directions. Development will be greatly affected by the neighbourhood effect and the transportation network. In addition, the topographical constraints on this development will also be important, especially when the urban areas of Sydney extend further outward from existing urban areas. Urban planning controls will play an important role in the process of urban development. The development may be more spread out if the strength of such planning controls is weak, or more concentrated if strong planning controls are in force. The prediction of the model has generated maps to show where the urban development of Sydney will occur in the two-and-a-half decades from 2006, which can be used to guide future urban development and land management.

6.5 CONCLUSION

This chapter presented results of the cellular automata model of urban development in Sydney based on fuzzy constrained transition rules. A retrospective approach in calibrating the model was applied. The calibration of the model was conducted systematically over time with the accuracies of the model's outcomes being compared to the actual urban development of Sydney between 1976 and 2006. Through this calibration, the impacts of various factors on the urban development of Sydney between 1976 and 2006 were evaluated. Although urban development can be affected by a number of factors, the self-propensity for development and neighbourhood support; transportation network; and topographical constraints such as slope, terrain, and coastal proximity attractions were identified as major drivers of Sydney's urban development. In addition, urban development in Sydney did not occur in all areas as was planned in the various planning programmes. However, such plannings affected Sydney's urban development by releasing the constraints of other factors such as slope of the area, and by reinforcing the impact of the transportation network.

The discussion on the model's outcomes under different neighbourhood sizes showed that, by varying the size of the neighbourhood configuration, significant impacts on the behaviour of the cellular automata model could be identified. For Sydney, a circular neighbourhood, the radius of which corresponds to a local block or neighbourhood community, was found to be the most accurate. When applying this model to simulate the process of urban development of another city, this neighbourhood size may not be applicable, and research will be needed to search for the appropriate neighbourhood size. What have not been studied in this model are the impacts of cell scale and the interactions between the cell scale and the neighbourhood size on the model's behaviour and outcomes. These need to be addressed in future research.

With the satisfactory calibration of the cellular automata model for the urban development of Sydney in a retrospective process, the model was applied to generate perspective views of the city for the next 25 years. Results presented realistic ideas of the expected patterns and composition of Sydney's urban development up to the year 2031. Although a number of statistics have been generated from these predictions, the value of the model was not so much its ability to predict the extent of urban areas but to identify areas with all the necessary conditions to support development. The perspective views of Sydney generated by the model highlighted the importance of urban planning controls, leading to the conservation of more non-urban land and the promotion of intensive development from the partly urban to the fully urban extent.

7 Future Research Directions

Modelling urban development has been the objective of urban research for over a hundred years (Batty 1998). This book presented a simulation model of urban development using the cellular automata approach incorporating fuzzy set theories and spatial information technology. Through the development of the model, the book contributes to the integration of fuzzy sets, cellular automata modelling, and geographical information systems (GIS) for urban development research. Significant advances have been made to the application of the fuzzy set theory for delimiting urban areas, as well as the application of fuzzy logic control in defining the transition rules of an urban cellular automata model. The application of the model has enabled an understanding to be developed of the controls and patterns of the urban development of Sydney from 1976 to 2006, and the provision of perspective views on the future directions of urban development under physical constraints, socio-economic conditions, and urban planning controls. Moreover, by implementing various transition rules in the model and examining its behaviour under various conditions, this book demonstrates how the cellular automata model of urban development can be used, first as an analytical tool to explore the rules or factors underpinning the process of urban development, and second as a planning tool to generate predictive scenarios of urban development and to answer various "what if" questions.

The principles of cellular automata was developed under the philosophy of open systems and the Chaos Theory, which claims that dynamic complex system behaviours emerge from local actions. By applying the cellular automata approach to model the process of urban development, the results obtained in this book demonstrate that the structure and behaviour of an urban system can be generated by locally defined transition rules. These transition rules can be represented by a set of simple "if–then" statements. The more one understands the system being modelled, the better the rules can be fine tuned for an accurate representation of the system.

This book demonstrates the power of using a fuzzy constrained cellular automata model and GIS technologies to simulate the process of urban development. As the model implements transition rules set in accordance with its calibration principles, the impact of each rule or factor on the model's behaviour and output can be identified and evaluated. Therefore, the model functions as an analytical tool to explore and evaluate the impact of rules or factors underpinning the process of urban development. With the flexibility of implementing transition rules, planners and decision makers can use the model to test various planning options to answer their "what if" questions. This modelling methodology contributes to the increasing literature on the application of cellular automata in urban planning as well as an understanding of the processes and controls of urban development.

The calibration of a simulation model is an essential component of model construction. A model is complete only if it has been tested and calibrated to be a true representation of the real system it models. In this book, the fuzzy constrained cellular automata model of urban development of Sydney was calibrated following a number of preset principles. The calibration was conducted using the actual urban extent data of Sydney between 1976 and 2006 in three consecutive ways: visual calibration, statistical calibration, and calibration over time. Through the simulation accuracy assessment, rules or factors controlling the process of urban development in Sydney were explored and evaluated. Predictions from the model have generated reasonable patterns for the urban development in Sydney for the two-and-a-half decades from the base year in 2006.

The fuzzy constrained cellular automata model developed in this book provides a useful tool for understanding the spatio-temporal processes of urban development in Metropolitan Sydney. However, the problems or limitations of using this approach in urban development modelling cannot be underestimated. This section addresses several limitations of the model and defines some future research directions for the application of the cellular automata modelling of urban development.

7.1   LOCAL AND GLOBAL TRANSITION RULES

One limitation of the cellular automata approach for modelling urban development is that its transition rules are defined by local scales, and the scale of the neighbourhood influencing the transition of cells applies uniformly to all cells over the whole area. In the real situation, not all factors affect urban development at a local scale. Some factors, such as urban planning and the transportation network, may affect urban development at a regional level. In particular, developments in information technology and telecommunications have had fundamental consequences for the patterns and processes of urban change throughout the world (Herbert and Thomas 1997). The extent of urban development as a locally defined process and the range of local and global factors being utilised in a cellular automata model of urban development need further research.

7.2   APPLICATIONS OF FUZZY SET AND FUZZY LOGIC

Previous applications of the cellular automata regard urban development as a binary process of non-urban to urban conversion (Clarke and Gaydos 1998; Wu 1998a,b,c, 1996; Wu and Webster 1998; Clarke, Hoppen, and Gaydos 1997). In practice, the boundary between non-urban and urban areas is not a sharp line. By introducing the fuzzy set theory for delimiting urban areas, a fuzzy membership grade was defined to denote the state of cells of a cellular automaton in an urban fuzzy set. With the membership grade, the state of all cells of the urban cellular automata were represented by a series of continuous values representing the fuzzy transition from non-urban to partly urban and then fully urban. Hence, there is no sharp boundary between cell states.

Many factors or rules, such as land topography, transportation, terrain, and coastal proximity affect urban development in a non-deterministic manner, characterising the process of development with a range of uncertainties. Recent studies on modelling urban development have addressed the non-deterministic uncertainties of urban development in a number of ways, including the application of stochastic and probability theories (for example, Bell, Dean, and Blake 2000; Li and Yeh 2000; Batty, Xie, and Sun 1999; White, Engelen, and Uljee 1997) and fuzzy sets (for example, Wu 1998b, 1996; Wang and Hall 1996; Wang 1994). Through the integration of cellular automata with the fuzzy set theory and fuzzy logic controls, the transition rules of urban cellular automata were implemented using a number of linguistic variables, such as "fast," "very fast," "normal," or "slow," "very slow," and so forth. The application of natural language statements made the modelling process more realistic and transparent.

With the application of fuzzy set and fuzzy logic, the cellular automata model of urban development constructed in this book has generated accurate outcomes in simulating the process of urban development in Sydney. However, the claim of urban development as a probabilistic and fuzzy process remains in the realm of assertion. Moreover, comparative analysis of the model's outcomes and those that might be generated by a standard cellular automata model might provide further information as to whether or not or to what extent the application of fuzzy sets has generated more accurate outcomes in modelling urban development.

The application of the fuzzy set theory to delimit urban areas provides a means to depict urban development as a continuous process. With the implementation of the fuzzy constrained transition rules of the urban cellular automata model, the process of urban development modelling is more realistic and transparent. However, as the membership grade of an urban fuzzy set and the linguistic modifiers of the transition rules were defined in a subjective manner, the interpretation of the model's results was restricted (Wu 1996). Other approaches, such as the artificial neural network (Openshaw and Openshaw 1997), the analytic hierarchy process (AHP; Saaty 1980), and the multicriteria evaluations (Wu and Webster 1998) were suggested to assist the definition of membership functions (Wu 1998c, 1996; Banai 1993). Applications of these approaches in the simulation of urban development based on the cellular automata approach are worthy of further research.

7.3 URBAN CONSOLIDATION AND ANTI-URBANISATION PROCESSES

This book focuses on simulating the process of urban expansion. No consideration was given to either the process of urban redevelopment or consolidation, or the urban decline or anti-urbanisation processes. With the rapid increase of urban population worldwide and the increased concerns on preserving prime agricultural land and the natural environment, the consolidated use of urban land becomes increasingly important. For Sydney, there has been a long debate on the redevelopment of existing urban areas (New South Wales Department of Planning 1995; Powell 1966). Worthy questions for further research concern whether or not the cellular automata model of urban development can be applied to simulate the process of urban redevelopment or the anti-urbanisation process, and how transition rules can be defined in an urban cellular automata model to simulate these processes.

7.4 THE SPATIAL AREA UNIT AND ITS INTERACTION WITH THE NEIGHBOURHOOD SCALE

When modelling spatial phenomena, a fundamental problem is the scale of unit to which the model is fitted (Fotheringham and Wong 1991). For urban models based on the cellular automata approach, this refers to both the size of cells and the neighbourhood scale. In this book, the cellular automata model of urban development was configured with a cell size of 250 m. The effects of different scales of neighbourhood were tested, and a circular neighbourhood with a radius reflecting the size of a local community was selected to model the urban development of Sydney. The theoretical basis for selecting the cell size and neighbourhood scale was unclear. The impact of different cell scales on the model's behaviour was not discussed; this requires further research. Furthermore, the relationships and interactions between the cell size and the neighbourhood scale on the behaviour of the urban cellular automata and its outcomes also require further research.

7.5 REAPPLICABILITY OF THE MODEL

The cellular automata model of urban development constructed in this book was applied and calibrated only for Metropolitan Sydney. Future work with this model will involve testing the model in other regions, such as in the highly urbanised city state of Singapore, or in the fast growing metropolitan region of Shanghai in China. Further applications of the model to predict and understand controls of urban growth will provide significant contributions to the cellular automata-based urban modelling as well as the understanding of urban growth itself.

References

Abel, D. J., Kilby, P. J. and Davis, J. R. 1994. The systems integration problem. *International Journal of Geographical Information Systems* 8: 1–12.

Allen, P. M. 1997. Cities and regions as evolutionary, complex systems. *Geographical Systems* 4: 103–30.

Allen, P. M. and Sanglier, M. 1978. Dynamic models of urban growth. *Journal of Social and Biological Structures* 1: 265–80.

Almeida, C. M. d., Batty, M., Monteiro, A. M. V., Câmara, G., Soares-Filho, B. S., Cerqueira, G. C. and Pennachin, C. L. 2003. Stochastic cellular automata modeling of urban land use dynamics: empirical development and estimation. *Computers, Environment and Urban Systems* 27: 481–509.

Alonso, W. 1964. *Location and land use*. Cambridge: Harvard University Press.

Altman, D. 1994. Fuzzy set theoretic approaches for handling imprecision in spatial analysis. *International Journal of Geographical Information Systems* 8: 271–89.

Angel, S. and Hyman, G. M. 1972. Urban transport expenditures. *Papers of Regional Sciences Association* 29: 105–23.

Anselin, L. and Getis, A. 1992. Spatial statistical analysis and geographical information systems. *Annals of Regional Science* 26: 19–33.

Anselin, L., Dodson, R. F. and Hudak, S. 1993. Linking GIS and spatial data analysis in practice. *Geographical Systems* 1: 2–23.

Aplin, G. 1982. Models of urban change: Sydney 1820–70. *Australian Geographical Studies* 20: 144–57.

Apostel, L. 1961. Towards the formal study of models in the non-formal sciences. In *The concept and the role of the model in mathematics and natural and social sciences, proceedings of the Colloquium sponsored by the Division of Philosophy of Sciences of the International Union of History and Philosophy of Sciences organized at Utrecht*, Ed. H. Freudenthal, 1–37. Dordrecht, Netherlands: D. Reidel.

Australian Bureau of Statistics. 2005. *Australian Standard Geographical Classification (ASGC)— Electronic Publication (1216.0)*. Canberra: Australian Bureau of Statistics. http://meteor. aihw.gov.au/content/index.phtml/itemId/312239. (accessed December 1, 2006).

Australian Bureau of Statistics. Census products 1996, 2001, 2006. Canberra: Australian Bureau of Statistics. http://www.abs.gov.au/websitedbs/d3310114.nsf/Home/census (accessed April 1, 2008).

Banai, R. 1993. Fuzziness in geographical information systems: contributions from the analytic hierarchy process. *International Journal of Geographical Information Systems* 7: 315–29.

Barnsley, M. and Barr, S. 1996. Inferring urban land-use from satellite sensor images using kernel-based spatial re-classification. *Photogrammetric Engineering and Remote Sensing* 62: 949–58.

Barnsley, M. and Barr, S. 1997. Distinguishing urban land-use categories in fine spatial resolution land-cover data using a graph-based, structural pattern recognition system. *Computers, Environment and Urban Systems* 21: 209–25.

Bassett, K. and Short, J. 1989. Development and diversity in urban geography. In *Horizons in human geography*, Eds. D, Gregory and R. Walford, 175–93. London: Macmillan.

Batty, M. 1976. *Urban modelling: algorithms, calibrations, predictions*. Cambridge, UK: Cambridge University Press.

Batty, M. 1981. Urban models. In *Quantitative geography: a British view*, Eds. N. Wrigley and R. J. Bennett, 181–91. London: Routledge and Kegan Paul.

Batty, M. 1982. The quest for the qualitative: new directions in planning theory and analysis. *Urban Policy and Research* 1: 15–23.

Batty, M. 1991. Generating urban forms from diffusive growth. *Environment and Planning A* 23: 511–44.

Batty, M. 1994. Using GIS for visual simulation modeling. *GIS World* 7: 46–8.

Batty, M. 1995. New ways of looking at cities. *Nature* 377: 574.

Batty, M. 1997. Cellular automata and urban form: a primer. *Journal of the American Planning Association* 63: 266–74.

Batty, M. 1998. Urban evolution on the desktop: simulation with the use of extended cellular automata. *Environment and Planning A* 30: 1943–67.

Batty, M. 2000. GeoComputation using cellular automata. In *GeoComputation,* Eds. S. Openshaw and R. J. Abrahart, 95–126. London: Taylor and Francis.

Batty, M. and Howes, D. 1996. Exploring urban development dynamics through visualization and animation. In *Innovation in GIS 3,* Ed. D. Parker, 149–61. London: Taylor and Francis.

Batty, M. and Longley, P. 1994. *Fractal cities: a geometry of form and function.* London: Academic Press.

Batty, M. and Xie, Y. 1994a. Modelling inside GIS: Part 1. Model structures, exploratory spatial data analysis and aggregation. *International Journal of Geographical Information Systems* 8: 291–307.

Batty, M. and Xie, Y. 1994b. Urban analysis in a GIS environment: population density modelling using ARC/INFO. In *Spatial Analysis and GIS*, Eds. S. Fotheringham and P. Rogerson, 189–220. London: Taylor and Francis.

Batty, M. and Xie, Y. 1994c. From cells to cities. *Environment and Planning B* 21: 31–48.

Batty, M. and Xie, Y. 1997. Possible urban automata. *Environment and Planning B* 24: 175–92.

Batty, M., Couclelis, H. and Eichen, M. 1997. Urban systems as cellular automata. *Environment and Planning B* 24: 159–164.

Batty, M., Longley, P. and Fotheringham, S. 1989. Urban growth and form: scaling, fractal geometry, and diffusion-limited aggregation. *Environment and Planning* 21: 1447–72.

Batty, M., Xie, Y. and Sun, Z. 1999. Modelling urban dynamics through GIS-based cellular automata. *Computers, Environment and Urban Systems* 23: 205–33.

Bell, M., Dean, C. and Blake, M. 2000. Forecasting the pattern of urban growth with PUP: a web-based model interfaced with GIS and 3D animation. *Computers, Environment and Urban Systems* 24: 559–81.

Benenson, I. 1998. Multi-agent simulations of residential dynamics in the city. *Computers, Environment and Urban Systems* 22: 25–42.

Benenson, I. 1999. Modeling Population Dynamics in the City: from a Regional to a Multi-Agent Approach. *Discrete Dynamics in Nature and Society* 3: 149–70.

Benenson, I. and Torrens, P. M. 2004. *Geosimulation: automata-based modeling of urban phenomena.* London: John Wiley.

Benenson, I., Omer, I. and Hatna, E. 2002. Entity-based modeling of urban residential dynamics—the case of Yaffo, Tel-Aviv. *Environment and Planning B: Planning & Design* 29: 491–512.

Berenji, H. R. 1992. Fuzzy logic controller. In *An Introduction to Fuzzy Logic Applications in Intelligent Systems*, Eds. R. R. Yager and L. A. Zadeh. London: Kluwer Academic.

Birkin, M., Clarke, G., Clarke, M. and Wilson, A. G. 1996. *Intelligent GIS: Location decisions and strategic planning.* Cambridge: GeoInformation International.

Bishop, Y. M. M., Fienber, S. E. and Holland, P. W. 1975. *Discrete multivariate analysis: theory and practice.* Cambridge: MIT Press.

Black, M. 1962. *Models and metaphors.* Ithaca, New York: Cornell University Press.

Bourne, L. S. 1971. *Internal structure of the city: readings on space and environment.* New York: Oxford University Press.

Bourne, L. S. 1982. *Internal structure of the city: readings on urban form, growth, and policy.* New York: Oxford University Press.

Boyce, D. E. 1965. The effect of direction and length of person trips on urban travel patterns. *Journal of Regional Science* 6: 65–80.

Bryant, C. R., Russwurm, L. H. and McLellan, A. G. 1982. *The city's countryside: land and its management in the rural–urban fringe.* New York: Longman.

Burgess, E. W. 1925. The growth of city: an introduction to a research project. In *The city,* Eds. R. E. Park, E. W. Burgess and R. D. McKenzie, 47–62. Chicago: The University of Chicago Press.

Burks, C. and Farmer, D. 1984. Towards modelling DNA sequences as automata. *Physica D* 10: 157.

Burrough, P. A. 1986. *Principles of geographic information systems for land resources assessment.* Oxford, U.K.: Clarendon Press.

Burrough, P. A. and Frank, A. U. 1995. Concepts and paradigms in spatial information: are current geographical information systems truly generic? *International Journal of Geographical Information Systems* 9: 101–16.

Caldwell, J. and Ram, Y. M. 1999. *Mathematical modelling: concepts and case studies.* Dordrecht, Netherlands, Kluwer Academic.

Caliper Corporation. 1983. *TransCAD: the premier GIS for transportation, logistics, and operations research.* http://www.caliper.com/default.htm (accessed May 5, 2008).

Campbell, J. B. 1996. *Introduction to Remote Sensing,* 2nd ed. New York: The Guilford Press.

Carter, H. 1995. *The study of urban geography,* 4th ed. London: Edward Arnold.

Cecchini, A. 1996. Urban modelling by means of cellular automata: generalised urban automata with the help on-line (AUGH) model. *Environment and Planning B* 23: 721–32.

Chapin, F. S. and Weiss, S. F. 1962a. *Urban growth dynamics in a regional cluster of cities.* New York: John Wiley.

Chapin, F. S. and Weiss, S. F. 1962b. Land development patterns and growth alternatives. In *Urban growth dynamics in a regional cluster of cities,* Eds. F. S. Chapin and S. F. Weiss, 425–58. New York: John Wiley.

Chapin, F. S. and Weiss, S. F. 1968. A probabilistic model for residential growth. *Transportation Research* 2: 375–390.

Chisholm, M. 1967. General systems theory and geography. *Transactions of the Institute of British Geographers* 42: 45–52.

Chorley, R. J. 1964. Geography and analogue theory. *Annals of the Association of American Geographers* 54: 127–37.

Chorley, R. J. and Haggett, P. 1967. *Models in geography.* London: Methuen.

Chorley, R. J. and Kennedy, B. A. 1971. *Physical geography: a systems approach.* London: Prentice Hall.

Chrisman, N. R. 1980. Assessing Landsat accuracy: a geographical application of misclassification analysis. *Second Colloquium on Quantitative and Theoretical Geography.* Cambridge: Trinity Hall.

Christaller, W. 1933 (translation 1966). *Central Places in Southern Germany.* Englewood Cliffs, New Jersey: Prentice Hall.

Claramunt, C. and Theriault, M. 1995. Managing time in GIS: an event-oriented approach. In *Recent Advances in Temporal Databases,* Eds. J. Clifford and A. Tuzhilin, 23–42. New York: Springer Verlag.

Clark, C. 1951. Urban population densities. *Journal of Royal Statistics Society, Serial A* 490–96.

Clarke, K. C. and Gaydos, L. J. 1998. Loose-coupling a cellular automaton model and GIS: long-term urban growth prediction for San Francisco and Washington/Baltimore. *International Journal of Geographical Information Sciences* 12: 699–714.

Clarke, K. C., Brass, J. A. and Riggan, P. 1995. A cellular automaton model of wildfire propagation and extinction. *Photogrammetric Engineering and Remote Sensing* 60: 1355–67.

Clarke, K. C., Hoppen, S. and Gaydos, L. J. 1997. A self-modifying cellular automaton model of historical urbanization in the San Franciso Bay area. *Environment and Planning B* 24: 247–61.

Clarke, M. 1990. Geographical information systems and model-based analysis. In *Geographical Information Systems for Urban and Regional Planning*, Eds. H. Scholten and S. Stillwell, 165–75. London: Kluwer Academic.

Cohen, J. 1960. A coefficient of agreement for nominal scales. *Educational and Psychological Measurement* 20: 37–40.

Cohen, K. J. and Cyert, R. M. 1961. Computer models in dynamic economics. *The Quarterly Journal of Economics* LXXV: 112–27.

Congalton, R. G. and Mead, R. A. 1983. A quantitative method to test for consistency and correctness in photo interpretation. *Photogrammetric Engineering and Remote Sensing* 49: 69–74.

Congalton, R. G., Oderwald, R. G. and Mead, R. A 1983. Assessing Landsat classification accuracy using discrete multivariate analysis statistical techniques. *Photogrammetric Engineering and Remote Sensing* 49: 1671–78.

Couclelis, H. 1985. Cellular worlds: a framework for modelling micro-macro dynamics. *Environment and Planning A* 17: 585–96.

Couclelis, H. 1989. Macrostructure and microbehavior in a metropolitan area. *Environment and Planning B* 16: 141–54.

Couclelis, H. 1997. From cellular automata to urban models: new principles for model development and implementation. *Environment and Planning B* 24: 165–74.

County of Cumberland Council. 1948. *The Planning Scheme for the County of Cumberland, New South Wales. The Report of the Cumberland County Council to the Hon. J. J. Cahill, M.L.A. Minister for Local Government.* Sydney: New South Wales Department of Planning.

Davidson, D., Theocharopoulos, S. and Bloksma, R. 1994. A land evaluation project in Greece using GIS and based on Boolean and fuzzy set methodologies. *International Journal of Geographical Information Systems* 8: 369–84.

Devroye, L., Gyorfi, L. and Lugosi, G. 1996. *A Probabilistic Theory of Pattern Recognition.* New York: Springer-Verlag.

Dickey, J. W. and Watts, T. M. 1978. *Analytic techniques in urban and regional planning: with applications in public administration and affairs.* New York: McGraw-Hill.

Dietzel, C. and Clarke, K. C. 2004. Spatial differences in multi-resolution urban automata modeling. *Transactions in GIS* 8: 479–92.

Dietzel, C. and Clarke, K. C. 2006. The effect of disaggregating land use categories in cellular automata during model calibration and forecasting. *Computers, Environment, and Urban Systems* 30: 78–101.

DiGregorio, S., Festa, D., Gattuso, D., Rongo, R., Spataro, W., Spezzano, G. and Vitetta, A. 1996. Cellular automata for freeway traffic simulation. In *Artificial worlds and urban studies,* Eds. E. Besussi and A. Cecchini, 365–92. Venice: DAEST.

Ding, Y. and Fotheringham, A. S. 1992. The integration of spatial analysis and GIS. *Computers, Environment and Urban Systems* 16: 3–19.

Doolen, G. and Montgomery, D. 1987. Magnetohydrodynamic cellular automata. *Physics Letters A* 120: 229.

Dragicevic, S. 2000. A fuzzy set approach for modelling time in GIS. *International Journal of Geographical Information Sciences* 14: 225–45.

Dragicevic, S. 2004. Coupling fuzzy sets theory and GIS-based cellular automata for land-use change modelling. In *Fuzzy Information, IEEE Annual Meeting of the Processing NAFIPS'04,* 203–07. Banff.

Dunn, E. S. 1954. *The location of agricultural production.* Gainesville, Florida: University of Florida Press.

Dyckman, J. W. 1963. The scientific world of the city planners. *American Behavioral Scientist* 6: 46–50.

Echenique, M. 1975. Urban development models: fifteen years of experience. In *Urban development models,* Eds. R. Baxter, M. Echenique, and J. Owers, 19–29. Hornby: The Construction Press.

Engelen, G., Geertman, S., Smits P. and Wessels C. 1999. Dynamic GIS and strategic physical planning: a practical application. In *Geographical Information Systems and Planning. Advances in Spatial Science,* Eds. J. Stilwell, S. Geertman and S. Openshaw, 87–111. Berlin: Springer.

Engelen, G., White, R. and Uljee, I. 1997. Integrating constrained cellular automata models, GIS and decision support tools for urban and regional planning and policy making. In *Decision support systems in urban planning,* Ed. H. Timmermans, 125–55. London: E. & F. N. Spon Ltd.

Engelen, G., White, R. and Uljee, I. 2001. Cellular automata as the core of integrated land-use models. In *Abstracting Integrated Spatial Models from Textual Scenarios,* Engelen, G., Straatman, B., Uljee, I. and Hagen, A. Report submitted to European Commission, DGXII, Science, Research and Sustainable Development, Brussels, Belgium.

Engelen, G., White, R., Uljee, I. and Drazan, P. 1995. Using cellular automata for integrated modelling of socio-environmental systems. *Environmental Monitoring and Assessment* 34: 203–14.

ESRI. 2004a. ArcGIS Desktop Help. Redlands, California: ESRI Press.

ESRI. 2004b. ArcGIS 9: ArcGIS Desktop Developer Guide. Redlands, California: ESRI Press.

Fischer, M. M. and Nijkamp, P. 1992. Geographical information systems and spatial analysis. *Annals of Regional Science* 26: 5–17.

Fischer, M. M., Scholten, H. J. and Unwin, D. 1996. *Spatial analytical perspectives on GIS.* London: Taylor and Francis.

Fitzgerald, S. 1987. *Rising damp: Sydney 1870–90.* Melbourne: Oxford University Press.

Flanagan, W. G. 1990. *Urban sociology: images and structure.* Sydney: Allyn and Bacon.

Forrester, J. W. 1961. *Industrial dynamics.* New York: John Wiley.

Forrester, J. W. 1969. *Urban dynamics.* Cambridge: The MIT Press.

Forrester, J. W. 1971. *World dynamics.* Cambridge: Wright-Allen Press.

Forster, B. 1983. Some urban measurements from Landsat data. *Photogrammetric Engineering and Remote Sensing* 49: 1693–707.

Forster, B. 1993. Coefficient of variation as a measure of urban spatial attributes using SPOT HRV and Landsat TM data. *International Journal of Remote Sensing* 14: 2403–09.

Fotheringham, A. S. and Rogerson, P. A. 1994. *Spatial Analysis and GIS.* London: Taylor and Francis.

Fotheringham, A. S. and Wong, D. W. S. 1991. The modifiable areal unit problem in multivariate statistical analysis. *Environment and Planning A* 23: 1025–44.

Fourastié, J. 1963. *Le Grand Espoir du XX Siécle.* Edition definitive. Paris: Gallimard.

Friedman, M. and Kandel, A. 1999. *Introduction to pattern recognition, statistical, structural, neural and fuzzy logic approaches.* Singapore: World Scientific.

Frost, L. and Dingle, T. 1995. Sustaining suburbia: An historical perspective on Australia's urban growth. In *Australian cities: strategies and policies for urban Australia in the 1990s,* Ed. P. Troy, 20–39. Melbourne: Cambridge University Press.

Fukunaga, K. 1990. *Introduction to Statistical Pattern Recognition,* 2nd ed. New York: Academic Press.

Gaines, B. R. 1975. *Multivalued logics and fuzzy reasoning.* Cambridge: BCS AISB Summer School.

Gaines, B. R. 1976. Foundations of fuzzy reasoning. *International Journal of Man–Machine Studies* 8: 623–68.

Gardner, M. 1972. The fantastic combinations of John Conway's new solitaire game 'Life'. *Scientific American* 233: 120–23.

Gazelton, N. W. J., Leahy, F. J. and Williamson, I. P. 1992. Integrating dynamic modelling with geographical information systems. *Journal of Urban and Regional Information Systems* 4: 47–58.

Geoscience Australia. 2006. GEODATA TOPO 250K Series 3. Canberra: Geoscience Australia. http://www.ga.gov.au/nmd/products/digidat/250k.htm (accessed May 1, 2007).

Goldstein, J. 1999. Emergence as a construct: history and issues. *Emergence: Complexity and Organization* 1: 49–72

Gong, P. and Howarth, P. 1990. The use of structural information for improving land-cover classification accuracies at the rural–urban fringe. *Photogrammetric Engineering and Remote Sensing* 56: 67–73.

Goodchild, M. F. 1992. Geographical information science. *International Journal of Geographical Information Systems* 6: 31–45.

Goodchild, M. F., Haining, R. and Wise, S. 1992. Integrating GIS and spatial data analysis: problems and possibilities. *International Journal of Geographical Information Systems* 6: 407–23.

Gose, E., Johnsonbaugh, R. and Jost, S. 1996. *Pattern recognition and image analysis*. Upper Saddle River, New Jersey: Prentice Hall.

Gottmann, J. 1961. *Megalopolis: the urbanized northeastern seaboard of the United States*. New York: 20th Century Fund.

Gryztzell, K. G. 1963. The demarcation of comparable city areas by means of population density [translated by Tom Fletcher]. *Lund studies in Geography, Series B, Human Geography*. No. 25. Lund, Sweden: Royal University of Lund, Department of Geography.

Guan, P. 1987. Cellular automaton public-key cryptosystem. *Complex Systems* 1: 51.

Hägerstrand, T. 1952. The propagation of innovation waves. *Lund studies in Geography, Series B, Human Geography*, 4: 16–17. Lund: Royal University of Lund, Department of Geography, Sweden.

Hägerstrand, T. 1965. A Monte Carlo approach to diffusion. *European Journal of Sociology* 6: 43–67.

Hägerstrand, T. 1967. *Innovation diffusion as a spatial process*. Chicago: University of Chicago Press.

Haggett, P. and Chorley, R. J. 1967. Models, paradigms and the new geography. In *Models in geography*, Eds. R. J. Chorley and P. Haggett, 19–42. London: Methuen.

Haines-Young, R. H. 1989. Modelling geographical knowledge. In *Remodelling geography*, Ed. W. Macmillan, 22–39. Oxford: Basil Blackwell.

Haines-Young, R. H. and Petch, J. H. 1986. *Physical geography: its nature and methods*. London: Harper & Row.

Hall, G., Wang, F. and Subaryono, G. 1992. Comparison of Boolean and fuzzy classification methods in land suitability analysis by using geographical information systems. *Environment and Planning A* 24: 497–516.

Hall, T. 1998. *Urban geography*. Routledge Contemporary Human Geography Series. London: Routledge.

Hambury, J. R. and Sharkey, R. H. 1961. *Land use forecast*. Chicago: Chicago Area Transportation Study.

Harris, B. and Batty, M. 1993. Locational models, geographical information and planning support systems. *Journal of Planning Education and Research* 12: 184–98.

Harris, C. D. and Ullman, E. L. 1945. The nature of cities. *Annals of the American Academy of Political and Social Sciences* 242: 7–17.

Harris, P. and Ventura, S. 1995. The integration of geographic data with remotely sensed imagery to improve classification in an urban area. *Photogrammetric Engineering and Remote Sensing* 61: 993–8.

Haslett, J., Wills, G. and Unwin, A. 1990. SPIDER—An interactive statistical tool for the analysis of spatially distributed data. *International Journal of Geographical Information Systems* 4: 185–96.

Haworth, R. J. 2003. The shaping of Sydney by its urban geology. *Quaternary International* 103: 41–55.

Hazelton, N. W. J., Leahy, F. J. and Williamson, I. P. 1992. Integrating dynamic modeling and geographic information systems. *Journal of the Urban and Regional Information Systems Association* 2: 47–58.

Henshall, J. D. 1967. Models of agricultural activity. In *Models in geography*, Eds. R. J. Chorley and P. Haggett, 425–58. London: Methuen.

Hepner, G., Houshmand, B., Kulikov, I. and Bryant, N. 1998. Investigation of the integration of AVIRIS and IFSAR for urban analysis. *Photogrammetric Engineering and Remote Sensing* 64: 813–20.

Herbert, D. T. and Thomas, C. J. 1997. *Cities in space: City as place,* 3rd ed. London: David Fulton.

Hesse, M. 1953. Models in physics. *British Journal of the Philosophy of Science* 4: 198–214.

Hillier, W. and Hanson, J. 1984. *The social logic of space*. Cambridge, U.K.: Cambridge University Press.

Hillis, D. 1984. The connection machine: a computer architecture based on cellular automata. *Physica D* 10: 213.

Holmblad, L. P. and Osterguard, J. J. 1981. Fuzzy logic control: operator experience applied in automatic process control. *ZKG International* 34: 127–33.

Holmblad, L. P. and Osterguard, J. J. 1982. Control of a cement kiln by fuzzy logic. In *Fuzzy Information and Decision Processes*, Eds. M. M. Gupta and E. Sanchez, 398–409. Amsterdam: North-Holland.

Hoover, E. M. 1936. The measurement of industrial location. *Review of Economics and Statistics* 18: 162–71.

Hoyt, H. 1939. *The structure and growth of residential neighborhoods in American cities*. Washington, D.C.: U.S. Government Printing Office.

Isard, W. 1956. *Location and space economy*. New York: John Wiley.

Itami, R. M. 1988. Cellular worlds: models for dynamic conceptions of landscape. *Landscape Architecture* 78: 52–7.

Itami, R. M. 1994. Simulating spatial dynamics: cellular automata theory. *Landscape and Urban Planning* 30: 27–47.

Itami, R. M. and Clark, J. D. 1992. Spatial dynamic simulations using discrete time and discrete event theory in cell based GIS systems. In *Proceedings: 5th International Symposium on Spatial Data Handling*, 702–12. Charleston, South Carolina.

Jakobson, L. and Prakash, V. 1971. *Urbanization and national development*. Beverly Hills, California: Sage.

Jantz, C.A., Goetz, S.J. and Shelley, M.K. 2004. Using the SLEUTH urban growth model to simulate the impacts of future policy scenarios on urban land use in the Baltimore-Washington metropolitan area. *Environment and Planning B: Planning and Design* 31: 251–71.

Jensen, J. R. 1996. *Introductory Digital Image Processing: a Remote Sensing Perspective*. Upper Saddle River, New Jersey: Prentice Hall.

John, R. I. 1995. Fuzzy Inferencing Systems—Problems and Some Solutions. Working Paper No. 62, Computing Sciences Research, School of Computing Sciences, De Montfort University. http://www.cse.dmu.ac.uk/~rij/newrep/newrep.html (accessed May 2, 2008).

Johnson, J. H. 1972. *Urban geography: an introductory analysis*. Oxford: Pergamon Press.

Johnston, R. J. and Wrigley, N. 1981. Urban geography. In *Quantitative geography: a British view*, Eds. N. Wrigley and R. J. Bennett, 335–51. London: Routledge and Kegan Paul.

Kickert, W. J. M. 1978. *Fuzzy theories on decisions-making: a critical review*. Boston: Leiden.

Kickert, W. J. M. and Mamdani, E. H. 1978. Analysis of a fuzzy logic controller. *Fuzzy Sets and Systems* 1: 29–44.

Kickert, W. J. M. and Van Nanta Lemka, R. H. 1976. Application of a fuzzy controller in a warm water plant. *Automatica* 12: 301–8.

Kilbridge, M. D., O'Block, R. P. and Teplitz, P. V. 1970. *Urban Analysis*. Boston: Harvard University.

Kocabas, V. and Dragicevic, S. 2006. Assessing cellular automata model behaviour using a sensitivity analysis approach. *Computers, Environment, and Urban Systems* 30: 921–53.

Kollias, V. and Viliotis, A. 1991. Fuzzy reasoning in the development of geographical information systems FRSIS: a prototype soil information system with fuzzy retrieval capabilities. *International Journal of Geographical Information Systems* 5: 209–23.

Kurtz, R. A. and Eicher, J. B. 1958. Fringe and suburb: a confusion of concepts. *Social Forces* 37: 32–7.

Langran, G. 1993. Issues of implementing a spatiotemporal system. *International Journal of Geographical Information Systems* 4: 305–14.

Langton, C. G. 1984. Self-reproduction in cellular automata. In *Cellular Automata*, Eds. D. Farmer, T. Toffoli and S. Wolfram, 135–44. Amsterdam: North-Holland.

Lathrop, G. T. and Hamburg, J. R. 1965. An opportunity-accessibility model allocating regional growth. *Journal of the American Institute of Planners* 31: 95–103.

Leão, S. Bishop, I. and Evans, D. 2004. Spatial-temporal model for demand allocation of waste landfills in growing urban regions. *Computers, Environment and Urban Systems* 28: 353–85.

Lee, C. C. 1990. Fuzzy Logic in Control Systems: Fuzzy Logic Controller, Part II. *IEEE Transactions on Systems, Man and Cybernetics* 20: 419–35.

Lee, D. B. 1973. Requiem for large-scale models. *Journal of the American Institute of Planners* 39: 163–78.

Levy, S. 1992. *Artificial Life*. New York: Vintage Books.

Li, X. and Yeh, A. G. 2000. Modelling sustainable urban development by the integration of constrained cellular automata and GIS. *International Journal of Geographical Information Systems* 14: 131–52.

Li, X. and Yeh, A. G. 2001. Calibration of cellular automata by using neural networks for the simulation of complex urban systems. *Environment and Planning A* 33: 1445–62.

Li, X. and Yeh, A. G. 2002. Integration of principal components analysis and cellular automata for spatial decision making and urban simulation. *Science in China (Earth Sciences)* 45: 521–9.

Lillesand, T. M. and Kiefer, R. W. 1994. *Remote Sensing and Image Interpretation*, 3rd ed. New York: John Wiley.

Linge, G. J. R. 1965. *The delimitation of urban boundaries for statistical purposes with special reference to Australia*. Canberra: Australia National University.

Liu, Y. and Phinn, S. R. 2003. Modelling urban development with cellular automata incorporating fuzzy-set approaches. *Computers, Environment, and Urban Systems* 27: 637–58.

Lo, C. P. 1997. Application of Landsat Thematic Mapper Data for quality of life assessment in an urban environment. *Computers, Environment and Urban Systems* 21: 259–76.

Lösch, A. 1943 (translation 1954). *The economics of location*. New Haven, Connecticut: Yale University Press.

Lowry, I. S. 1964. *A model of metropoles*. Santa Monica, California: The Rand Corporation.

Lowry, I. S. 1965. A short course in model design. *Journal of the American Institute of Planners* 31: 158–66.

Ma, Z. K. and Redmond, R. L. 1995. Tau coefficients for accuracy assessment of classification of remote sensing data. *Photogrammetric Engineering and Remote Sensing* 61: 435–9.

Macrae, N. 1992. *John von Neumann.* New York: Pantheon.

Mainster, M. 1992. Cellular automata: retinal cells, circulation, and patterns. *Eye* 6: 420.

Makins, M. (Ed.) 1995. *Collins English Dictionary.* 3rd ed. updated. Glasgow: Harper Collins.

Malm, R., Olsson, G. and Wärneryd, O. 1966. Approaches to simulations of urban growth. *Geografiska Annaler* 48B: 9–22.

Mamdani, E. and Assilian, S. 1975. An experiment in linguistic synthesis with a fuzzy logic controller. *International Journal of Man–Machine Studies* 7: 1–13.

Mandelas, E. A., Hatzichristos, T. and Prastacos, P. 2007. A Fuzzy Cellular Automata Based Shell for Modeling Urban Growth—A Pilot Application in Mesogia Area. In *10th AGILE International Conference on Geographic Information Science 2007, Denmark.*

March, L. 1974. Quantifying the environment. In *Models, Evaluation and Information Systems for Planners, LUBFS Conference Proceedings No. 1,* Eds. J. Perraton and R. Baxter. Lancaster: MTC Construction.

McNeill, D. and Freiberger, P. 1994. *Fuzzy logic.* New York: Simon and Schuster.

Meadows, P. 1957. Models, system and science. *American Sociological Review* 22: 3–9.

Menard, A. and Marceau, D. J. 2005. Exploration of spatial scale sensitivity in geographic cellular automata. *Environment and Planning B* 32: 693–714.

Mesev, V. 1998. The use of census data in image classification. *Photogrammetric Engineering and Remote Sensing* 64: 431–8.

Mesev, V., Longley, P. Batty, M. and Xie, Y. 1995. Morphology from imagery: detecting and measuring the density of urban land-use. *Environment and Planning A* 27: 759–80.

Miller, H. J. 1991. Modeling accessibility using space-time prism concepts within geographic information systems. *International Journal of Geographical Information Systems* 5: 287–301.

Moller-Jensen, L. 1990. Knowledge based classification of urban area using texture and context information in Landsat TM imagery. *Photogrammetric Engineering and Remote Sensing* 56: 899–904.

Morrill, R. L. 1965. *Migration and the spread and growth of urban settlement.* Lund studies in geography, Series B, Human Geography, 26. Lund: Gleerup.

Nagel, K., Rasmussen, S. and Barrett, C. 1997. Network traffic as a self-organized critical phenomenon. In *Self-organization of complex structures: from individual to collective dynamics,* Eds. F. Schweitzer and H. Haken, 579–92. New York: CRC Press.

Naylor, T. H. and Finger, J. M. 1967. Verification of computer simulation models. *Management Science* 14:B92–B106.

NCGIA (National Center for Geographic Information and Analysis). 1995. [Final Version]. Advancing geographic information science: a proposal to NSF by NCGIA. http://www.ncgia.ucsb.edu/secure/main.html. (accessed May 2, 2008).

New South Wales Department of Environment and Planning. 1988. *Sydney into its Third Century: Metropolitan Strategy for the Sydney Region.* Sydney.

New South Wales Department of Planning. 1992. *Updating the Metropolitan Strategy.* Sydney.

New South Wales Department of Planning. 1993a. *Sydney's Future: A Discussion Paper on Planning the Greater Metropolitan Region.* Sydney.

New South Wales Department of Planning. 1993b. *Integrated Transport Strategy for Greater Sydney.* Sydney.

New South Wales Department of Planning. 1995. *Cities for the 21st Century: Integrated Urban Management for Sydney, Newcastle, the Central Coast and Wollongong.* Sydney.

New South Wales Department of Planning. 2005. *City of cities: a plan for Sydney's future.* Sydney.

New South Wales Planning and Environment Commission. 1980. *Review: Sydney Region Outline Plan.* Sydney.

New South Wales State Planning Authority. 1968. *Sydney Region Outline Plan (1970–2000 A.D.): A Strategy for Development.* Sydney.

Norlén, U. 1975. *Simulation model building, a statistical approach to modeling in the social sciences with the simulation method.* New York: John Wiley.

Nyerges, T. L. 1995. Geographical information system support for urban/regional transportation analysis. In *The Geography of Urban Transportation,* 2nd ed. Ed. S. Hanson, 240–68. New York: Guildford.

O'Sullivan, D. 2001. Exploring spatial process dynamics using irregular cellular automaton models. *Geographical Analysis* 33: 1–18.

Openshaw, S. 1984. The modifiable areal unit problem. *Concepts and Techniques in Modern Geography.* Norwich: Geo Books.

Openshaw, S. and Openshaw, C. 1997. *Artificial intelligence in geography.* New York: John Wiley.

Peuquet, D. J. 1988. Representations of geographic space: towards a conceptual synthesis. *Annals of the Association of American Geographers* 78: 375–94.

Peuquet, D. J. 1994. It's about time: a conceptual framework for the representation of temporal dynamics in geographic information systems. *Annals of the Association of American Geographers* 3: 441–61.

Peuquet, D. J. and Duan, N. 1995. An event-based spatiotemporal data model (ESTDM) for temporal analysis of geographical data. *International Journal of Geographical Information Systems* 1: 7–24.

Phinn, S. R., Stanford, M., Scarth, P., Murray, A. T. and Shyy, P. T. 2002. Monitoring the composition of urban environments based on the vegetation-impervious surface-soil (VIS) model by subpixel analysis techniques. *International Journal of Remote Sensing* 23: 4131–53.

Portugali, J. and Benenson, I. 1995. Artificial planning experience by means of a heuristic cell-space model: simulating international migration in the urban process. *Environment and Planning A* 27: 1647–65.

Powell, A. J. 1966. Sydney: Redevelopment in a metropolitan region context. In *Urban redevelopment in Australia-papers presented to a joint urban seminar held at the Australia National University October and December 1966.* Ed. P. N. Troy, Research School of Sciences, Urban Research Unit, Australia National University, Canberra.

Power, C., Simms, A. and White, R. 2001. Hierarchical fuzzy pattern matching for the regional comparison of land use maps. *International Journal of Geographical Information Science* 15: 77–100.

Prigogine, I. and Stengers, I. 1984. *Order out of chaos: man's new dialogue with nature.* London, UK: Heinemann.

Pryor, R. J. 1968. Defining the rural–urban fringe. *Social Forces* 47: 202–15.

Putnam, S. 1992. *Integrated Urban Models 2.* London: Pion Press.

Raper, J. and Livingstone, D. 1995. Development of a geomorphological spatial model using object-oriented design. *International Journal of Geographical Information Systems* 9: 359–83.

Reichenbach, H. 1951. *The rise of scientific philosophy.* Berkeley: University of California Press.

Roberts A. 1978. Planning Sydney's Transport 1875–1900. In *Nineteenth-Century Sydney: Essays in urban history,* Ed. M. Kelly, Sydney: Sydney University Press.

Robinson, G. M. 1998. *Methods and techniques in human geography.* New York: John Wiley.

Rosenfield, G. H. 1986. Analysis of thematic map classification error matrices. *Photogrammetric Engineering and Remote Sensing* 52: 681–6.

Rosenfield, G. H. and Fitzpatrick-Lins, K. 1986. A coefficient of agreement as a measure of thematic classification accuracy. *Photogrammetric Engineering and Remote Sensing* 52: 223–7.

Saaty, T. L. 1980. *The analytic hierarchy process: planning, priority setting, resource allocation.* New York: McGraw-Hill.

Samat, N. 2006. Characterizing the scale sensitivity of the cellular automata simulated urban growth: a case study of the Seberang Perai Region, Penang State, Malaysia. *Computers, Environment and Urban Systems* 30: 905–20.

Sander, L. 1986. Fractal growth processes. *Nature* 322: 789.

Shevky, E. and Bell, W. 1955. *Social area analysis.* Stanford: California, Stanford University Press.

Shevky, E. and Williams, M. 1949. *The social area of Los Angeles: analysis and typology.* Berkeley: University of California Press.

Shi, W. and Pang, M.Y.C. 2000. Development of Voronoi-based cellular automata—an integrated dynamic model for Geographical Information Systems. *International Journal of Geographical Information Science* 14: 455–74.

Silva, E.A. and Clarke, K.C. 2002. Calibration of the SLEUTH urban growth model for Lisbon and Porto, Portugal. *Computers, Environment and Urban Systems* 26: 525–52.

Silverman, B. W. 1986. *Density estimation for statistics and data analysis.* New York: Chapman and Hall.

Snodgrass, R. T. 1992. Temporal databases. In *Theory and Methods of Spatio-Temporal Reasoning in Geographic Space*, Eds. A. U. Frank, I. Campari, and U. Formentini, 22–64. Springer.

Spearritt, P. 2000. *Sydney's century: a history.* Sydney: University of New South Wales Press.

Spearritt, P. and DeMarco, C. 1988. *Planning Sydney's Future.* Sydney: Allen & Unwin.

Statistics Canada. 1996. *About 1996 Census Tables on the Internet.* Statistics Canada. http://www.statcan.ca/english/census96/about.html (accessed May 2, 2008).

Stevens, D. and Dragićević, S. 2007. A GIS-based irregular cellular automata model of land-use change. *Environment and Planning B: Planning and Design* 34: 708–24.

Story, M. and Congalton, R. G. 1986. Accuracy assessment, a user's perspective. *Photogrammetric Engineering and Remote Sensing* 52: 397–9.

Sui, D. Z. 1998. GIS-based urban modelling: practices, problems, and prospects. *International Journal of Geographical Information Sciences* 12: 651–71.

Sui, D. Z. and Zeng, H. 2001. Modeling the dynamics of landscape structure in Asia's emerging desakota regions: a case study in Shenzhen. *Landscape and Urban Planning* 53: 37–52.

Syphard, A. D., Clarke, K. C. and Franklin, J. 2005. Using a cellular automaton model to forecast the effects of urban growth on habitat pattern in southern California. *Ecological Complexity* 2: 185–203.

Takeyama, M. and Couclelis, H. 1997. Map dynamics: integrating cellular automata and GIS through Geo-Algebra. *International Journal Geographical Information Sciences* 11: 73–91.

Theobald, D. M. and Hobbs, N. T. 1998. Forecasting rural land-use change: a comparison of regression- and spatial transition-based models. *Geographical and Environmental Modelling* 2: 65–82.

Theodoridis, S. and Koutroumbas, K. 1999. *Pattern recognition.* San Diego: Academic Press.

Thomas, R. W. and Huggett, R. J. 1980. *Modelling in geography: a mathematical approach.* New Hersey: Barnes & Noble Books.

Tobler, W. R. 1970. A computer movie simulating urban growth in the Detroit region. *Economic Geography* 26: 234–40.

Tobler, W. R. 1979. Cellular Geography. In *Philosophy in Geography*, Eds. S. Gale and G. Olsson, 379–86. Dordrecht, Netherlands: D. Reidel.

Toffoli, T. and Margolus, N. 1987. *Cellular Automata Machines.* Cambridge: The MIT Press.

Torrens, P. M. 2000. *How cellular models of urban systems work (1. theory).* Centre for Advances Spatial Analysis Working Paper 28. University College London, UK. http://www.casa.ucl.ac.uk/working_papers.htm (accessed May 2, 2008).

Torrens, P. M. 2003. Automata-based models of urban systems. In *Advanced Spatial Analysis: the CASA Book of GIS*, Eds. P. Longley and M. Batty, 61–81. Redlands, California: ESRI Press.

Torrens, P. M. and Benenson, I. 2005. Geographic Automata Systems. *International Journal of Geographical Information Science*, 19: 385–412.

U.S. Bureau of the Census. 1960. *U.S. Census of Population 1960 United States Summary*, IA: xiii.

Ulam, S. 1976. *Adventures of a Mathematician*. New York: Charles Scribner's Sons.

Uljee, I., Engelen, G. and White, R. 1996. Ramco Demo Guide. Workdocument CZM-C 96.08. The Hague: Coastal Zone Management Centre, National Institute for Coastal and Marine Management.

Umbers, I. G. and King, P. J. 1980. An analysis of human decision-making in cement kiln control and the implications for automation. *International Journal of Man–Machine Studies* 12: 11–23.

United Nations. 2006. *World Urbanization Prospects: the 2005 Revision*. Population Division, Department of Economic and Social Affairs, United Nations. http://www.un.org/esa/population/publications/WUP2005/2005wup.htm (accessed May 2, 2008).

Vichniac, G. 1984. Simulating physics with cellular automata. *Physica D* 10: 96.

Vincent, J. 1986. Cellular Automata: A model for the formation of color patterns in molluscs. *Journal of Molluscan Studies* 52: 97.

Vitousek, P. M. 1994. Beyond global warming: ecology and global change. *Ecology* 75: 1861–76.

von Bertalanffy, L. 1968. *General systems theory: foundations, development, applications*. New York: George Braziler.

von Neumann, J. 1966. *The Theory of Self-reproducing Automata*, Ed. A. Burks. Urbana, Illinois, University of Illinois Press.

von Thünen, J. H. 1826. *Der Isolierte Staat* (English translation by C. M. Wartenberg, edited by P. Hall, *von Thünen's Isolated State*, 1966. Oxford: Pergamon Press).

Wagner, D. F. 1997. Cellular automata and geographic information systems. *Environment and Planning B* 24: 219–34.

Wang, F. 1994. Towards a natural language user interface: an approach of fuzzy query. *International Journal of Geographical Information Systems* 8: 143–62.

Wang, F. and Hall, G. B. 1996. Fuzzy representation of geographical boundaries in GIS. *International Journal of Geographical Information Systems* 10: 573–90.

Ward, D. P., Murray, A. T. and Phinn, S. R. 2000. Monitoring growth in rapidly urbanising areas using remotely sensed data. *Professional Geographer* 52: 371–86.

Webb, A. 1999. *Statistical pattern recognition*. New York: Oxford University Press.

Weber, A. 1909. *Über den Standort der Industrien*. Tübingen: Reine Theorie des Standort.

Webster's. 1964. *Third new international dictionary of the English language [unabridged]*. Springfield, Massachusetts: G. & C. Merriam Company.

Wehrwein, G. S. 1942. The rural–urban fringe. *Economic Geography* 18: 217–28.

Wenstøp, F. 1976. Deductive verbal models of organizations. *International Journal of Man–Machine Studies* 8: 293–311.

Wheeler, J. O. 1970. Transport inputs and residential rent theory: an empirical analysis. *Geographical Analysis* 2: 43–54.

White, R. 1998. Cities and cellular automata. *Discrete Dynamics in Nature and Society* 2: 111–25.

White, R. and Engelen, G. 1993. Cellular automata and fractal urban form: a cellular modelling approach to the evolution of urban land-use. *Environment and Planning A* 25: 1175–99.

White, R. and Engelen, G. 1994. Cellular dynamics and GIS: modelling spatial complexity. *Geographical Systems* 1: 237–53.

White, R. and Engelen, G. 1997. Cellular automata as the basis of integrated dynamic regional modeling. *Environment and Planning B* 24: 235–46.

White, R. and Engelen, G. 2000. High-resolution integrated modelling of the spatial dynamics of urban and regional systems. *Computers, Environment and Urban Systems* 24: 383–400.

White, R., Engelen, G. and Uljee, I. 1997. The use of constrained cellular automata for high-resolution modelling of urban land-use dynamics. *Environment and Planning B* 24: 323–43.

Whitesitt, J. E. 1961. *Boolean algebra and its applications. Reading,* Massachusetts: Addison-Wesley.

Wilson, A. G. 1970. *Entropy in urban and regional modeling.* London: Pion.

Wilson, A. G. 1981a. Catastrophe theory and bifurcation. In *Quantitative geography: a British view*, Eds. N. Wrigley and R. J. Bennett, 192–201. London: Routledge and Kegan Paul.

Wilson, A. G. 1981b. *Catastrophe theory and bifurcation: applications to urban and regional systems.* London: Croom Helm.

Wilson, A. G. 1984. One man's quantitative geography: frameworks, evaluations, uses and prospects. In *Recollections of a revolution: geography as spatial science*, Eds. M. Billinge, D. Gregory and R. L. Martin, 200–26. London: Macmillan.

Wilson, A. G. 1989. Classics, modelling and critical theory: human geography as structured pluralism. In *Remodelling geography*, Ed. W. Macmillan, 61–9. Oxford: Basil Blackwell.

Wingo, L. 1961. *Transportation and urban land.* Baltimore, MD: John Hopkins University Press.

Winston, D. 1957. *Sydney's great experiment: the progress of the Cumberland County Plan.* Sydney: Halstead Press.

Witten, T. A. and Sander, L. M. 1981. Diffusion-limited aggregation: a kinetic critical phenomenon. *Physical Review Letters* 47: 1400–03.

Wolfram, S. 1983. Statistical mechanics of cellular automata. *Reviews of Modern Physics* 55: 601–44.

Wolfram, S. 1984. Cellular automata as models of complexity. *Nature.* 311: 419–24.

Wolfram, S. 1994. *Cellular automata and complexity: collected papers.* Reading, MA: Addison-Wesley.

Wolfram, S. 2002. *A New Kind of Science.* Champaign, Illinois: Wolfram Media.

Woodcock, C. E. and Gopal, S. 2000. Fuzzy set theory and thematic maps: accuracy assessment and area estimation. *International Journal of Geographical Information Science* 14: 153–72.

Wu, F. 1996. A linguistic cellular automata simulation approach for sustainable land development in a fast growing region. *Computers, Environment, and Urban Systems* 20: 367–87.

Wu, F. 1998a. An experiment on the generic polycentricity of urban growth in a cellular automatic city. *Environment and Planning B: Planning and Design* 25: 731–52.

Wu, F. 1998b. Simulating urban encroachment on rural land with fuzzy-logic-controlled cellular automata in a geographical information system. *Journal of Environmental Management* 53: 293–308.

Wu, F. 1998c. SimLand: a prototype to simulate land conversion through the integrated GIS and CA with AHP-derived transition rules. *International Journal of Geographical Information Science* 12. 63–82.

Wu, F. 2002. Calibration of stochastic cellular automata: the application to rural–urban land conversions. *International Journal of Geographical Information Science* 16: 795–818.

Wu, F. and Webster. C. J. 1998. Simulation of land development through the integration of cellular automata and multicriteria evaluation. *Environment and Planning B: Planning and Design* 25: 103–26.

Wu, F. and Webster. C. J. 2000. Simulating artificial cities in a GIS environment: urban growth under alternative regulation regimes. *International Journal of Geographical Information Science* 14: 625–48.

Yang, X. and Lo, C. P. 2003. Modelling urban growth and landscape change in the Atlanta metropolitan area. *International Journal of Geographical Information Science* 17: 463–88.

Yeh, A. G. and Li, X. 2006. Error and uncertainties in urban cellular automata. *Computers, Environment, and Urban Systems* 30: 10–28.

Zadeh, L. A. 1962. From circuit theory to systems theory. *IRE Proceedings* 50: 856–65.

Zadeh, L. A. 1965. Fuzzy sets. *Information and Control* 8: 335–53.

Zadeh, L. A. 1971. Towards a theory of fuzzy systems. In *Aspects of Network and Systems Theory*, Eds. R. E. Kalman and N. DeClaris. Winston: Holt Rinehart.

Zadeh, L. A. 1975a. The concept of a linguistic variable and its application to approximate reasoning—I. *Information Sciences* 8: 199–249.

Zadeh, L. A. 1975b. The concept of a linguistic variable and its application to approximate reasoning—II. *Information Sciences* 8: 301–57.

Zadeh, L. A. 1975c. The concept of a linguistic variable and its application to approximate reasoning—III. *Information Sciences* 9: 43–80.

Zimmermann, H. J. 1985. *Fuzzy set theory and its applications.* Dordrecht, Netherlands, Kluwer-Nijhoff.

Zimmermann, H. J. 1987. *Fuzzy sets, decision making, and expert systems.* Boston: Kluwer Academic.

Zimmermann, H. J. 1991. *Fuzzy set theory and its applications,* 2nd revised ed. London: Kluwer Academic.

Index

A

Abstraction levels, 4, 7
Accelerating factors for development, 80, 81;
    See also Transportation
Action-at-a-distance neighbourhood, 41–42
Agent-based models, 17, 18–19
Aggregation
    diffusion-limited (DLA), 17–18
    spatial data, 22
Aggregative approach, modelling, 11
Agricultural location, Von Thünen model, 4, 7,
    8, 11
Analogue models, 4
Analytical capabilities, 20
Analytical hierarchy process (AHP), 48,
    121, 161
Analytic models, 5
AND function, 58
Anti-urbanisation processes, future research
    directions, 161
Approximations of reality, 3
Approximations of unreality, 7
ArcCatalog, 130
ArcGIS
    cellular automata modeling, 128
    Spatial Analyst extension, 117
    Sydney, 101, 129–130
    TDI point and line density index
        computation, 117
ArcMacro Language (AML), 128, 129
ArcMap, 130
Area data sets, Sydney model calibration
    with, 142
Area definition, urban; *See* Urban area definition
Area units, spatial, 22, 162
Artificial neural network-based models,
    49–50, 161
Atlanta, 39
Attractive effects, cell weight, 43
Automata, 17, 18; *See also* Cellular automata

B

Behavioural approach
    cell-based framework, 36
    models and modelling theory, 13–14
Behaviour of models, 21
Behaviour of systems, 27

Behaviour patterns
    behavioural approach to modelling, 13
    cellular automata versus multiple agent
        systems, 18–19
Bifurcation, 16
Binary cell states, cellular automata, 41
Binary map data, weights of evidence
    approach, 50
Binary states, cellular automata models, 52
Boolean logic, 55, 58, 64, 66, 69
Boundaries/boundary representations
    cellular automata models, 41, 52
    fuzzy constrained cellular automata model,
        54–55
    fuzzy systems, 19, 60
    iCity model, 41
    urban area definition
        as fuzzy process, 60
        Sydney, 102, 104, 105–107
Boundary representations, 54
Burgess, E.W., 8, 9

C

C (programming language), 129
Cadastral boundaries, 54
Cadastral parcels, 41
Cadastral units, 40
Calibration, 160
    cell size in ArcGIS GUI, 130
    fuzzy transition rules, 81–82
    model building stages, 6
    SLEUTH model phases, 47
    Sydney model, 103, 158
        GIS implementation, 131, 132
        neighbourhood scale impact on model's
            results, 151
        planning effects, 118–119
        principles, 120–122
        results under different neighbourhood
            scales, 146
        sequence of, 133, 134, 135
        simulation accuracy measurement,
            122–128
        topographically constrained
            development, 142
        transportation supported development, 143
        urban planning schemes, 144
    transportation density index (TDI) and, 116
    urban area definition, 60

Cartographic time, 51
Cascading systems, 14, 15
Cell-by-cell comparisons, error matrix
    approach, 122–123, 124
Cells
    cellular automata, 25, 28
    Sydney model, 158
Cell size
    graphic user interface design, 130, 131
    Sydney model, 111–112, 158
Cell state
    fuzzy transition rules, 52
    Sydney model specification, 111–112
Cellular automata, 21, 25–52, 159, 160
    contemporary practices, 38–51
        from binary and multiple to continuous cell
            states, 41
        modelling time, 51
        neighbourhood definitions, 41–45
        space tessellation, 38–41
        transition rule variations, 45–51
    fuzzy constrained; *See* Fuzzy constrained
        cellular automata model
    fuzzy set theory and fuzzy logic controls, 161
    modelling
        complex features of, 29–30
        game, 25–27
        simple model, 27–28
    multiple agent systems (MAS) versus, 18–19
    principles of, 5
    SimLand components, 48
    simple model, 28, 29
    Sydney; *See also* Sydney, modelling
        GIS implementation, 128–129
        study area, 87
        urban planning schemes, 103
    urban modelling, 30–38
        advantages of, 33–35
        early applications of, 35–38
        example of, 30–33
Cellular automata machine (CAM), 35
"Cellular Geography" (Tobler), 36
Census data
    defining urban areas, 59, 61, 85, 102, 107
    Sydney
        physical urban areas, 102
        visualisation of changes, 107, 108, 109
Central Place Theory (Christaller), 8
Centre of area method, defuzzification, 82
Change dynamics, constrained cellular automata
    model, 38, 39
Chaos, cellular automata features, 30
Chaos Theory, 5, 16–17, 21
Chapin, Stuart, 13, 14, 36
Chicago Area Transportation Study (CATS), 11
Chicago School of Human Ecology, 9
China, Guanzhou City, 35, 39, 42, 47–48

Cincinnati, 39
Circular neighbourhoods, 44, 45
Cities as self-organizing systems, 16–19
Cities for the 21st Century (1995), 97, 118,
    119, 144
City of Cities (2005), 97–99, 119
City of Cumberland Planning Scheme (1948),
    90–92
Clark, C., 12
Classical logic, 64–65
Classical models of urban ecology, 10
Coastal proximity, 115, 119, 142, 158, 160
Cognitive behavioural approach to modelling,
    13–14
Coincidence matrices, 44
Combined transition rules, 45
Commission error; *See* Errors, omission and
    commission
Community services, Sydney model, 119, 145
Compatibility/incompatibility, cell weight, 43
Complement, fuzzy operation, 58, 59
Complex behaviour
    cellular automata
        transition rule variations, 45
        urban modelling, 32–33
    local actions and, 159
Complexity
    cellular automata features, 29–30
    mechanisms of, 27
Complexity theory, 22, 27
Complex open systems, 8
Component object model (COM)-compliant
    programming languages, 129, 130
Computation errors, 21
Computation time
    iCity model, 41
    resolution selection, 39
Computer technology
    custom-built software, 129
    factor analysis, 10
    and mathematical models, 8
Concentric Zone Model (Burgess), 8
Concentric Zone Model of Burgess, 9
Conceptual models, 4
Confidence level, rule firing threshold, 78
Consolidation, future research directions, 161
Constrained cellular automata model, 39
    development of, 38
    fuzzy; *See also* Fuzzy constrained cellular
        automata model
    Sydney; *See* Sydney, modelling
    testing, 39
    transition rule variations, 45–46, 47–48
Constraining factors for development
    as fuzzy variables, 80–81
    terrain; *See* Topography

Constraints on action, 8–9
Consumers, neoclassical approach to modelling, 11–12
Contemporary practices of urban development modelling, 16–20
    cities as self-organizing systems, 16–19
    fuzzy set and fuzzy logic, 19
    GIS and, 19–20
Continuous cell states, 41
Contraposition, 65
Controls, 14, 15
    behavioural approach to modelling, 13
    local versus global, 42
Controls, fuzzy constrained cellular automata-based modelling, 62–70
    calibration of, 82
    linguistic variables and fuzzy logic, 63–67
        basic logic terms and reasoning, 64–65
        fuzzy logic, 66–67
        linguistic variables, 63–64
    processes, 67–68
    Sydney model, perspective views under different planning control factors, 153–157
Conway, John, 26, 27, 36
Correlations, modelling pitfalls, 7
County of Cumberland Region (CCR), 85–88
Crisp set theory, 55
Cross-tabulation map, 44
Cycle, growth, 46–47

D

Data
    census; See Census data
    GIS sources, 8, 128–129
    remote sensing; See Remote sensing technology
    Sydney model calibration with, 142
    technical problems in modelling, 22
Database (transaction) time, 51
Decision-making process
    behavioural approach to modelling, 13–14, 36
    fuzzy logic control, 68
    fuzzy systems, 19
Defuzzification, 19, 68
    fuzzy constrained cellular automata model, 82, 83
    modified error matrix approach, 125–126
Dendritic growth, 18
Descriptive models, 5
Deterministic models, 5
Detroit region, 36
Diagrammatic models, 4

Diffusion, SLEUTH model growth coefficients, 46–47
Diffusion-limited aggregation (DLA), 17–18, 45
Diffusion models, 36
Digital computing; See Computer technology
Distance-decay effects, 42, 43, 44
Distortion, neighbourhood type and, 43–44
Dynamic conceptualisation of space, 22
Dynamic models, 5
Dynamics of change, constrained cellular automata model, 38, 39
Dynamic spatial development, cellular automata, 35
Dynamic systems, 15
    cellular automata features, 30
    cellular automata technique applications, 27
    fuzzy representation of boundaries, 54
    local actions and, 159
DYNAMO (computer language), 15

E

Ecological approach, models and modelling theory, 9–10
Ecology, factorial, 10
Economic activities, theoretical approaches to modelling, 8
Economic equilibrium approach, 12
Elementary cellular automata, 27
Emergent properties, 30, 33
Employment, neoclassical approach to modelling, 12
Entropy-maximising spatial interaction model, 11
Equilibrium, theory of, 11, 12
Error matrix approach, 122–123, 124
    kappa coefficient analysis, 126–128
    modified, 124–126, 127
Errors, omission and commission
    Sydney model, 140, 141, 142
        topographically constrained development, 142
        transportation supported development, 144
        transportation-supported development, 144
Errors in computation, 21
Evidence, weights of, 50
Evolutionary and complex systems, 8
Evolution of system, cellular automata applications, 27
    What if experiments, 34–35
Excluded areas, Sydney, 102–103
Experimentation
    model applications, 3
    Sydney model, 146
Exponential fuzzy membership functions, 57

F

Factor analysis, 10, 14–15, 121
Factorial ecology, urban, 21
Feedback loops, Systems Dynamics
   technique, 15
Flowchart, Sydney model simulation and
   calibration sequence, 134, 135
Fluctuations, order from, 17
Forecasting; *See* Prediction/predictive nature
   of models
Formulation, model building stages, 6
Fractals, 16, 21
Fringe, rural-urban, 54
Functionally dependent model, Tobler, 37
Future development, Sydney, 152–153
Future planning; *See* Planning
Future prospects
   models and modelling, 22–23
   research directions, 159–162
Fuzzification, 19, 62, 65, 67, 68
Fuzzy constrained cellular automata model,
   53–83, 159, 160, 161
   developing for urban modelling, 70–83
      defuzzification, 82, 83
      fuzzy transition rules and inferencing,
         76–82
      speed of development as fuzzy set, 73–76
      temporal process, 70–73
   fuzzy logic control in cellular automata-based
      modelling, 62–70
      control, 67–68
      control in cellular automata, 69–70
      linguistic variables and fuzzy logic, 63–67
   Sydney; *See* Sydney, modelling
   transition rule variations, 47–48
   urban development and fuzzy sets, 53–62
      fuzzy set theory, 55–59
      geographical boundary representations,
         54–55
      urban development as fuzzy process,
         59–62
Fuzzy logic, 65
   future research directions, 160–161
   models and modelling practices, 19
Fuzzy logic control; *See* Controls, fuzzy
      constrained cellular automata-based
      modelling
Fuzzy rule transition mechanisms, 52
Fuzzy sets
   future research directions, 160–161
   models and modelling practices, 19
Fuzzy set theory, 16, 41, 161
   cellular automata, 52
   fuzzy constrained cellular automata model,
      55–59
      definition of fuzzy set, 55–56

   fuzzy operation, 58–59
   membership function, 56–58
Fuzzy logic, 65
Sydney urban area definition with, 104–109
   boundary definition, 105–107
   urban area criteria for statistical purposes,
      104–105
   visualising development in space and time,
      107–108, 109
   urban area definition, 59–60
FuzzyUrbanCA, 130
Fuzzy values, model variables, 6–7

G

Game of Life (Conway), 26, 27, 36, 69
Gaming style, cellular automata modeling, 48
Gardiner, M., 27, 69
General models, 5
General Systems Theory, 14, 16
GEODATA TOPO-250K, 100–101, 103
Geographical boundary representations, fuzzy
      constrained cellular automata model,
      54–55
Geographical information systems (GIS),
      23, 159
   ANN training data, 49–50
   cellular automata, advantages of, 34–35
   modelling problems, 22
   modelling techniques, 8, 19–20
   SimLand components, 48
   Sydney
      ArcGIS approach, 129–130
      cellular automata modelling and GIS,
         128–129
      graphic user interface design, 130–131
      model calibration, 131, 132
      physical urban areas, 102
      visualising development in space and
         time, 107
Geographical model, 36
Geographical model, Tobler, 37
GIS; *See* Geographical information systems
      (GIS)
Global behaviour of self organising systems,
      locally defined transition rules, 42
Global changes, local behaviour and, 17, 26
Global controls, 42
Global factors, 160
Goodness-of-fit, 121
Graphic user interface design, ArcGIS, 129,
      130–131
Gravity model, 10–11
GRID environment, 129
Group decisions, behavioural approach to
      modelling, 14

Growth characteristics
  diffusion-limited aggregation, 17–18
  SLEUTH model, 46–47
  Systems Dynamics technique, 15
  unconstrained, and Sydney model, 139–142
Growth coefficients, SLEUTH model, 46–47
Growth trends, 1
Guanzhou City, China, 35, 39, 42, 47–48

H

Hägerstrand, Torsten, 36
Harris, C.D., 8, 9, 10
Height of land; *See* Topography
Heuristic control rules, 67
Historical data sources, 128
Historical model, Tobler, 37
History, development stages, 70, 71
Holiday settlements, Sydney, 105
Houston, 39
Hoyt, H., 8, 9
Human choices, 8–9, 13–14

I

iCity model, 41, 44, 128
Iconic models, 4
If-then format, 159
  fuzzy constrained cellular automata model
    development, 77
  fuzzy logic control, 67, 68, 69
IF-THEN statements, 28, 30, 31–33, 34; *See also*
    Transition rules
Implementation, model building stages, 6
Incompatible states, cell weight, 43
Independent model, Tobler, 37
Individuals, behavioural approach to modelling,
    13–14, 36
Industrial Location (Weber), 8
Inference engine, 68
Inferencing, fuzzy transition rules, 76–82
  primary transition rules, 76–77
  rule calibration, 81–82
  rule firing threshold, 77–78
  secondary transition rules, 79–81
Information loss, defuzzification, 82
Information technology, neighbourhood size, 43
Infrastructure
  Sydney model, 119, 145
  transportation; *See* Transportation
Initial state, cellular automata features, 30, 33–34
Innovation-diffusion models, 14, 35–36
Intersection, fuzzy operation, 58
Irregular neighbourhoods, 44
Irregular spatial units, advantages of cellular
    automata, 38–41

Isolated State, von Thünen model
Iterative process, modelling as, 7

J

Java, 129

K

Kappa coefficient analysis, 126–128; *See also*
    Simulation accuracy, Sydney model
Knowledge base, fuzzy logic control, 67–68

L

Land cover, cellular automata modelling, 41
Land excluded from urban development, Sydney,
    102–103
Land features; *See* Boundary representations;
    Slope; Topography
Landsat imagery, 39, 60
Land use
  cellular automata
    data, 41
    What If experiments, 34–35
  constrained cellular automata model, 38, 39
  remote sensing technology, 59, 60
  theoretical approaches to modelling, 8
  Tobler's model, 36
  urban area definition, 61
Language, natural; *See also* Linguistic variables
  cellular automata modeling, 48
  model types, 4
Lattice network, 25
Law of Gravitation, Newton, 10
Layers, map, 22
Less developed countries, 1, 21
Limited-map metaphor, 22
Linear analysis, modelling techniques, 8
Linear membership functions
  defuzzification, 82, 83
  fuzzy sets, 56, 57
Linear programming, 14–15
Line density index, TDI, 116, 117
Lines, feature representation as, 22
Linguistic cellular automata model, 42, 47–48
Linguistic variables, 161
  and fuzzy logic, 63–67
    basic logic terms and reasoning, 64–65
    transition rules, 69–70
Linguistic variables, fuzzy systems, 19
Local action
  and complex system behavior, 159
  global behaviour effects, 17, 26, 42
Local factors, 160
Locally defined process, 160

Locally defined transitions, 22, 42
Local rules, mathematical pattern generation,
     25–26
Local scale, cellular automata features, 30
Local settlement network modelling, 36
Location costs, neoclassical approach to
     modelling, 12
Logarithmic fuzzy membership functions, 57
Logic
   classical, 64–65
   fuzzy, 19, 58, 69–70, 160–161; *See also* Fuzzy
        constrained cellular automata model
Logistic curve of urban development, 70, 71,
     72, 73
Lösch, A., 4, 11

M

Macros, GIS, 20
Macroscale factors, cellular automata modeling, 48
Macroscale models, 11, 45
Majority state of cells, 126
Maps
   layers, 22
   remote sensing data, 60
   simulation accuracy measurement, 122–128
   thematic, 54
   weights of evidence approach, 50
Mathematical methods
   linguistic modifiers, 63–64
   model building, 6, 8
   simple cellular automata, 29
   systems approach, 14–15
   transition rule configuration, 48–49
Mathematical models, 4, 5
   artificial neural network-based models, 49–50
   digital computing and, 8
   fuzzy logic and, 67
   transition rules derived from, 48
Mathematical programming, 8
Max criterion method, defuzzification, 82
Maximisation rule of utility, 11, 12, 13, 21
Maximum entropy law, 11
Mean of maximum method, defuzzification, 82
Mean value of factors, 48–49
Membership function, 161
   defuzzification, 82, 83
   fuzzy boundary definition, 105–106
   fuzzy constrained cellular automata model
        development, 73
     rule firing threshold, 77–78
     speed of development as fuzzy set, 74–76
   fuzzy logical operations, 58
   fuzzy logic control, 67
   fuzzy set theory, 58–59
   urban area definition, 60, 61–62

Membership value, Sydney urban fuzzy set, 111
Metaphors of urban growth, cellular automata
     as, 38
Metropolitan Strategic Plan (2005), 118, 144,
     153–157
Microscale factors, cellular automata
     modeling, 48
Microscale model, cellular automata, 45
Migration, modelling, 36
Milwaukee, 39
Minimum confidence level, rule firing
     threshold, 78
Mobility, residential, 14
Modelling time, cellular automata, 51
Models and modelling, 2–7
   characteristics of models, 3–4
   contemporary practices of, 16–20
     cities as self-organizing systems, 16–19
     fuzzy set and fuzzy logic, 19
     GIS and, 19–20
   definitions, 2
   need for models, 2–3
   pitfalls, 7
   problems and prospects, 20–23
     future prospects, 22–23
     technical problems, 22
     theoretical problems, 20–21
   procedures of model building, 6–7
   theoretical approaches to, 7–16
     behavioural approach, 13–14
     neoclassical approach, 11–13
     social physical approach, 10–11
     systems approach, 14–16
     urban ecological approach, 9–10
   types of models, 4–6
Modifiable area unit problem (MAUP), 38–39
Modified error matrix approach, 124–126, 127
Modus ponens/modus tollens, 65, 66
Moore Neighbourhoods, 28, 31, 41, 43
More developed countries, model
     applicability, 21
Morphological systems, 14, 15
Motivations, behavioural approach to modelling,
     13–14
Moving window, 48–49
Multicriteria analysis, 14–15, 48, 121, 161
Multiple agent systems (MAS), 18–19
Multiple cell states, cellular automata, 41
Multiple land uses, 50
Multiple Nuclei Model (Harris and Ullman), 8
Multiple Nuclei Model of Harris and Ullman,
     9–10
Multiple regression analysis, 48–49
Multivariate framework, stochastic cellular
     automata model, 50
Multivariate model, Tobler, 37

N

Natural language; *See* Language, natural;
    Linguistic variables
Natural urban growth
    Sydney model transition rules, 151
    unconstrained, 139
Nearest-neighbour values, cellular automata
        transition rules, 27
Negative effects, cell weight, 43
Neighbourhood definitions
    action-at-a-distance neighbourhood, 41–42
    ArcGIS graphic user interface design, 131
    cellular automata, 41–45
    irregular, 44
    and model performance and outcome, 40
    sensitivity analysis, 44–45
    size, 42–43
    type, 43–44
Neighbourhood effect, diffusion models, 36
Neighbourhoods
    cellular automata, 52
        elements of, 28
        features of, 30
        versus multiple agent systems, 18–19
        sensitivity analysis, 44–45
        transition rules, 27
    graphic user interface design, 130
    interaction with spatial area unit, future
            research directions, 162
    Sydney model
        and model calibration, 151
        perspective views under different planning
            control factors, 157
        results under different scales, 146–148
        simulation accuracies of model over time,
            149–151
        size of, 158
        specification, 112
    von Neumann, 26
Neighbourhood support, Sydney model,
        144, 158
Neoclassical approach, 11–13, 20–21
.NET, 129
Neural networks; *See* Artificial neural network-
        based models
New spreading centre growth, SLEUTH model,
        46–47
Newton, Isaac, 10
Nonlinear programming, 14–15
Nonlinear systems
    model principles, 5
    systems approach, 16
Normative models, 4, 5
Nuclei, urban ecological approach, 9–10
Numerical values, model building stages, 6

O

Objectives, classifications of models, 5, 6
Object-oriented programming languages, 129
Omission error; *See* Errors, omission and
        commission
One-dimensional cellular automata, 27
Open systems, 16, 30, 159
Open systems theory, 17
Operators, logical, 58, 64, 65
Order, from fluctuations, 17
OR function, 58
Outcomes, model building stages, 6, 7
Overlap, crisp set theory, 55

P

Parameters
    estimation of, 121
    model building stages, 6, 7
Partial models, 5
Partly urban state, 62
    Sydney, 107, 108, 111
    transition rules, 114
Pattern recognition, 124
Physical approach, models and modelling
        theory, 10–11
Physical urban areas, Sydney, data collection
        and processing, 102
Planning
    global controls, 42
    Sydney
        1948: City of Cumberland Planning
            Scheme, 90–92
        1968: Sydney Region Outline Plan, 93, 94
        1988: Sydney into Its Third Century,
            95, 96
        1995: Cities for the 21st Century, 97
        2005: City of Cities, 97–99
        data collection and processing, 103
    Sydney model transition rules, 151, 158
        constrained development by secondary
            transition rules, 118–119
        impact of individual factors on
            development, 138
        impact on development, 118–119, 144–145
        perspective views on development under
            different control factors, 153–157
        simulation accuracy under different
            transition rules in 2006, 138–139, 140
        What if experiments, 34–35
Point density index, TDI, 116
Points, feature representation as, 22
Polygons
    feature representation as, 22
    irregular spatial units, 44

Population
  cellular automata, macroscale model, 45
  fuzzy logic control, 67–68
  Sydney
    fuzzy boundary definition, 106, 107
    historical development, 88, 90
    physical urban areas, 102
    urban area criteria for statistical purposes,
      104, 105
  theoretical approaches to modelling, 8
  urban area definition, 59–60, 61
Prediction/predictive nature of models, 4, 5, 38
  accuracy, and validity of model, 121
  SLEUTH model phases, 47
  Sydney model, 158
  systems approach, 16
  unpredictable uncertainty, 17
  What If experiments, 34–35, 120, 146, 159
Principal component analysis (PCA), 14–15, 48,
    49, 121
Probabilistic models, 5, 50–51
Probabilistic process, urban development as, 54
Probability, suitability based components
    analysis model, 49
Problem specification, model building, 6
Process-response systems, 14, 15
Producers, neoclassical approach to modelling,
    11–12
Programming, systems approach, 14–15
Programming languages, 129
Proximity, Tobler's rule, 36
Python, 129

Q

Qualitative analysis, 8
Quality of data, 22
Quantitative revolution in geography, 8

R

Raster cells, 22
Raster GIS, 35, 50, 116, 128, 129
Real-world time, 51
Reapplicability of models
  future research directions, 162–163
  and value of models, 4
Reasoning, 64–65
Rectangular neighbourhoods, 43–44
Regional context, fuzzy set definition, 55–56
Regression methods, 121
Relief, 31–33; See also Topography
Remote sensing technology, 22
  GIS data, 128
  land use change simulation, 39
  urban area definition, 59, 60, 61

Research, model applications, 3
Residential land development, Wingo's theory
    of, 11–12
Residential mobility, 14
Resolution
  cellular automata
    calibration at different scales, 40
    irregular spatial units, 41
  computation time considerations, 39
  spatial unit cells, 38–41
Retrospective prediction approach, 121
Road-influenced growth, SLEUTH model,
    46–47
Rule-firing threshold, 115
  defined, 78
  primary transition rules, 113
  Sydney model, 144, 145, 153, 154
Rule of least costs, 11
Rule of thumb experience, fuzzy logic
    control, 67
Rules; See also Transition rules
  fuzzy constrained cellular automata model
    development, 70, 77–78
  fuzzy logic control, 67, 68
  Game of Life, 26, 27
Rural-urban fringe, 54

S

San Francisco Bay region, 34–35, 40, 47
Satellite data; See Remote sensing technology
Scale
  cellular automata, 48
    calibration at different scales, 40
    features of, 30
    resolution selection, computation time
      considerations, 39
    sensitivity analysis, 44–45
    space tessellation, 38
    Sydney neighbourhood scale impact on
      model's results, 146–148
Scale models, 4
Secondary transition rules
  ArcGIS approach, 131
  Sydney model, 114–119, 135
    planning schemes, 118–119
    topographical conditions, 114–115
    transportation network, 116–117, 118
    urban planning schemes, 144
Second law of thermodynamics, 11
Sector Model (Hoyt), 8
Sector model of Hoyt, 9
Self-organizing paradigm, 22
Self-organizing systems
  cellular automata features, 30
  cities as, 16–19

Self-propensity
  flexibility in rule implementation, 119
  Sydney model, 144, 151, 158
Self-replicating systems, 25, 26
Self-similar pattern generation, cellular
    automata features, 30
Sensitivity analysis, neighbourhood definitions,
    44–45
Sensitivity of models, 21
Service accessibility, Sydney model, 119, 145
Set membership function, 56–57
Settlement network modelling, 36
SimLand, 48
Simulation
  modelling techniques, 8
  Sydney model, 133
Simulation accuracy, Sydney model, 121, 138
  ArcGIS calibration, 131
  under different transition rules in 2006,
    138–139
  individual factors, impact on development,
    138–142
  measurement of, 122–128
    error matrix approach, 122–123, 124
    kappa coefficient analysis, 126–128
    modified error matrix approach,
      124–126, 127
  model calibration, 131
  neighbourhood scale impact on model's
    results, 149–151
  overall results under all transition rules,
    134–137, 138
  over time, 149–151
  planning effects, 144
  results under different neighborhood scales,
    146–148
  transportation supported development, 143
Simulation models, 5, 134
Site-specific accuracy, error matrix approach,
    122–127
SLEUTH model, 46–47
Slope; See also Topography
  constraining factors, fuzzy inferencing,
    80, 81
  locally specified processes, 42
  SLEUTH model growth coefficients, 47
Social area model, 21
Socialist countries, model applicability, 21
Social physical approach, 20
  cellular automata applications, 35–36
  models and modelling theory, 10–11
Social systems, modelling, 3
Soft computing, 16
Software, custom-built, 129
Software development kit (SDK), ArcGIS, 129
Space, dynamic conceptualisation of, 22

Space tessellation, 35, 38–41
  elements of, 28
  scale of, 111
Spatial accuracy, Sydney model, 140, 141
Spatial Analyst extension, ArcGIS, 117
Spatial area units, 22, 162
Spatial cells, cellular automata, 25
Spatial data
  aggregation, 22
  Sydney model, 145
Spatial dimensions, cellular automata, 25–26
Spatial dynamics, cellular automata
  advantages of, 34–35
  continuous cell state use, 41
Spatial ecology, 27
Spatial interaction model,
    entropy-maximising, 11
Spatial patterns, 2
  dynamic processes, 17
  neoclassical approach to modelling, 11
  social physical approach, 35–36
Spatial scale, cellular automata, 44–45
Spatial units
  calibration at different scales, 40
  interaction with neighbourhood scale, future
    research directions, 162
  space tessellation, 38–41
    irregular spatial units, 40–41
    regular cells of small or large resolution,
      38–40
  Sydney, 102
    data collection and processing, 102
    urban area definition with fuzzy set theory,
      107–108, 109
Speed of development as fuzzy set, 73–76, 80
  flexibility in rule implementation, 119–120
  fuzzy constrained cellular automata model,
    73–76
  rule calibration, 81–82
  Sydney
    parameter specification, 113
    secondary transition rules, 117, 118
Spontaneous urban growth, 46
SPOT, 60
Sprawl, Sydney, 90
Spreading centre growth, SLEUTH model,
    46–47
Stability characteristics, Systems Dynamics
    technique, 15
Stages, model outcome generation in, 5
State, cellular automaton elements, 28
State changes; See Transition rules
State of cell, simple cellular automata model, 27
State of system, simple cellular automata
    model, 27
Static models, 5

Statistical calibration, 122
Statistical estimation, stochastic cellular
     automata model, 50–51
Statistical measures, model validation, 121
Statistical mechanics, 11
Stochastic models, 5, 50–51
Strength parameter, planning effects on Sydney
     development, 119
Structure
   cellular automata features, 30
   urban, theories of, 12
Suggestive nature, models, 3–4
Suitability based components analysis model, 49
Supply-demand relationships, 11
Sydney, modelling
   calibration, 120–128
     principles, 120–122
     simulation accuracy measurement, 122–128
   GIS implementation, 128–132
     ArcGIS approach, 129–130
     cellular automata modelling and GIS,
       128–129
     graphic user interface design, 130–131
     model calibration, 131, 132
   individual factors, impact on development,
     138–146
     other transition rules, 145–146
     planning schemes and policies, 144–145
     topographically constrained
       development, 142
     transportation-supported development,
       142–144
     unconstrained urban growth, 139–142
   neighbourhood scale impact on model's
     results, 146–151
     and model calibration, 151
     results under different scales, 146–148
     simulation accuracies of model over time,
       149–151
   perspective views on development to year
     2031, 151–157
     under different planning control factors,
       153–157
     Metropolitan Strategic Plan (2005) impact
       on future development, 153
     transportation infrastructure improvement
       and future development, 152–153
   specification, 111–120
     cell size and state, 111–112
     neighbourhood configuration, 112
     temporal dimension, 120
     transition rules, 113–120
   summary of results from model, 133–138
     overall results under all transition rules,
       134–137, 138
     simulation and calibration sequence,
       133, 134

Sydney, urban development and visualisation,
     85–110
   data collection and processing, 100–103
     land excluded from urban development,
       102–103
     physical urban areas, 102
     topographic data, 100–101
     transportation network, 101
     urban planning schemes, 103
   spatial concepts, 86
   study area, 86
   urban area definition with fuzzy set theory,
     104–109
     boundary definition, 105–107
     urban area criteria for statistical purposes,
       104–105
     visualising development in space and time,
       107–108, 109
   urban development and planning, 85–100
     1948: City of Cumberland Planning
       Scheme, 90–92
     1968: Sydney Region Outline Plan, 93, 94
     1988: Sydney into Its Third Century, 95, 96
     1995: Cities for the 21st Century, 97
     2005: City of Cities, 97–99
     historical threads of development, 88–90
Sydney into Its Third Century (1988), 95, 96,
     118, 119, 144
Sydney Region Outline Plan (1968), 93, 94, 118
Syllogisms, 64, 65
Systems analysis, 14–16
Systems behaviour, simple cellular automata
     model, 27
Systems Dynamics, 15

T

Tautologies, 64, 65
TDI (transportation density index), 116–117
Telecommunications, 43
Temporal processes; See Time/temporal
     dimension
Terminology
   basic logic terms, 64–65
   fuzzy representation of boundaries, 54–55
Terrain, 160; See also Slope; Topography
   cellular automata, 31–33
   topographically constrained development, 142
Tessellation, space, 28, 35, 38–41, 111
Testing of models, 21
Thematic map, 54
Theoretical approaches to urban development
     modelling, 7–16
   behavioural approach, 13–14
   neoclassical approach, 11–13
   social physical approach, 10–11

systems approach, 14–16
urban ecological approach, 9–10
Theoretical problems, models and modelling, 20–21
Theoretical quantitative geography, 36
Theory-based models, 5
Theory-laden models, 5
Thermodynamics, second law of, 11
Threshold values, urban area definition, 59–60, 61, 62
Time/temporal dimension
  cellular automata, 25, 52
    elements of, 28
    modelling, 51
  classifications of models, 5
  conceptualisation levels, 22
  fuzzy constrained cellular automata model, 70–73
  fuzzy representation of boundaries, 55
  fuzzy set definition, 56
  simple cellular automata model, 27
  Sydney model, 146
    specification, 120
    transportation infrastructure improvement, 153
    visualising development in space and time, 107–108, 109
  systems approach, 16
Tobler, Waldo, 36
Tobler's models, 36, 37
Tobler's rule, 36
Topic-neutral items, logic, 64
Topography, 160
  Sydney, data collection and processing, 100–101
  Sydney model, 142, 158
    constrained development by secondary transition rules, 114–115
    impact of individual factors on development, 138
    perspective views under different planning control factors, 157
    simulation accuracy under different transition rules in 2006, 138–139, 140
    topographically constrained development, 142
    transition rules, 151
    transportation-supported development, 142–143
    transportation supported development, 142–143
    urban planning schemes, 144, 145
  urban modelling, 31–33
Transaction time, 51
TransCAD, 20
Transition functions, artificial neural network-based models, 49–50

Transition potentials, constrained cellular automata, 46
Transition probability, moving window, 48
Transition rules, 159, 161
  cellular automata, 27, 52
    combined, 45
    elements of, 28
    features, 30
    fuzzy constrained, 47
    mathematical methods, 48–49
    mathematical representation, 28
    neighbourhood size and, 43
    simple, 27
    urban modelling, 30, 31–33, 34
  cellular automata, variations in rules, 45–51
    artificial neural network-based models, 49–50
    constrained cellular automata, 45–46
    fuzzy constrained cellular automata model, 47–48
    rules derived from other models, 48–49
    SLEUTH model, 46–47
    stochastic models, 50–51
  fuzzy constrained cellular automata development, 76–82
    primary, 76–77
    rule calibration, 81–82
    rule firing threshold, 77–78
    secondary, 79–81
  local and global, future research directions, 160
  locally specified processes, 150
  Sydney model
    ArcGIS graphic user interface design, 131
    constrained development by secondary rules, 114–119
    flexibility in implementation, 119–120
    miscellaneous, 145–146
    natural growth controlled by primary transition rules, 113–114
    and overall results, 134–137, 138
    overall results under all transition rules, 134–137, 138
    simulation accuracy under different transition rules in 2006, 138–139, 140
    simulation and calibration sequence, 134
    specification, 113–120
    transportation supported development, 142
    unconstrained urban growth, 139
    urban planning schemes, 144
Transition zones, fuzzy representation of boundaries, 54
Transportation, 160
  accelerating factors, fuzzy inferencing, 81
  cellular automata
    constrained, 46
    SLEUTH model, 46–47
    urban modelling, 32, 33

infrastructure improvements, effects on future development, 152–153
locally specified processes, 42
neighbourhood size, 42
Sydney
  data collection and processing, 101
  early planning, 90
  historical development, 88, 89
Sydney model, 145, 158
  impact of individual factors on development, 138, 142–144
  perspective views under different planning control factors, 157
  secondary transition rules, 116–117, 118
  simulation accuracy under different transition rules in 2006, 138–139, 140
  transition rules, 151
  urban planning schemes, 144, 145
  theoretical approaches to modelling, 8
Transportation costs, neoclassical approach to modelling, 11, 12
Transportation density index (TDI), 116–117
Truth values
  fuzzy logic, 66
  logic, 64, 65
Turbulence, cellular automata features, 30
Turing, Alan, 25
Type, neighbourhood definitions, 43–44

U

Ulam, Stanislaw, 25
Ullman, E.L., 8, 9, 10
Uncertainty, unpredictable, 17
Unconstrained urban growth, and Sydney model, 139–142
Uniform conditions assumption, 30–31
Union, fuzzy operation, 58, 59
Unpredictable uncertainty, 17
Updating time, 51
Urban area definition
  with fuzzy set theory, 104–109
    boundary definition, 105–107
    urban area criteria for statistical purposes, 104–105
    visualising development in space and time, 107–108, 109
  Sydney
    data collection and processing, 102
    visualising development in space and time, 107, 108, 109
  urban development as fuzzy process, 59–60
Urban centres, neoclassical approach to modelling, 13
Urban development as fuzzy process, 59–62

Urban ecological approach, 9–10
Urban factorial ecology, 21
Urban planning; See Planning
Utility maximisation rule, 11, 12, 13, 21

V

Vacant cells, transition rule variations, 45
Validation, 120; See also Calibration
  model building stages, 7
  modelling pitfalls, 7
Value system, behavioural approach to modelling, 13
Vector structure, iCity model, 41
Verbal models, 4
Verbal values, model variables, 6–7
Verification of models, 21, 120; See also Calibration
Visual Basic 6, 129
Visual Basic for Applications (VBA), 129, 130
Visual C++, 129
Visual calibration, 122
Visualisation
  GIS platform and, 22
  Sydney urban area changes, 107–108, 109
Vocabulary, logic, 64
von Bertalanffy, L., 14
von Neumann, John, 25
von Neumann Neighbourhoods, 26, 41
von Thünen model, 4, 7, 8, 11, 79
Voronoi-based cellular automata model, 40–41, 44

W

Washington/Baltimore region, 35, 40, 47
Weber, A., 8, 11
Weight, cell, 43
  multiple regression method for computation, 48
Weights of evidence, stochastic cellular automata model, 50
What If scenarios, 120, 159
  cellular automata, advantages of, 34–35
  Sydney model, 146
Window, calibration, ArcGIS model, 121, 130
Window, moving, 48–49
Wolfram, Stephen, 27, 28

Z

Zones of transition, fuzzy representation of boundaries, 54
Zoning, SLEUTH model, 47

Milton Keynes UK
Ingram Content Group UK Ltd.
UKHW040057071024
449327UK00019B/614

9 780367 577438